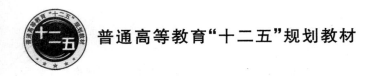

普通高等教育"十二五"规划教材

A Guide to the Problems in University Physics
(Second Edition)

大学物理双语课程解题指导书(第2版)

张晓光　芦鹏飞　席丽霞
韩利红　张　勇　编著

北京邮电大学出版社
www.buptpress.com

内容简介

本书由北京邮电大学几位资深物理教师编写,他们在于2004年开始的北京邮电大学与英国伦敦大学玛丽女王学院学士学位联合培养项目中执教大学物理双语课程多年,具有丰富的教学经验。

大学物理双语教学是大学物理课程现代化的一个改革方向,是高等教育与国际接轨的一项尝试。国内外的大学物理课程在广度和深度方面不尽相同,这是由中西方的教学模式和理念均有很大的不同所致。为了与课程教学相适应,学生们在掌握物理学的基本理论基础之上,还需要通过一定数量、一定难度的题目训练,达到巩固所学知识和提高运用能力的目的。由于受大多数英文原版教材的深度所限,题目的难度与中文教程的题目难度相比有所不足,因此有必要编写一套涵盖选择、填空和计算的大学物理双语教学习题指导教材,有效地补充解题指导这一教学环节,可以作为教师上习题课以及学生课后自习的参考教材。

本书的第1版自出版以来,受到了广大读者的欢迎。本版对于部分章节进行了较大幅度的改写,重新选择了一定量的题目。

图书在版编目(CIP)数据

大学物理双语课程解题指导书 / 张晓光等编著. - - 2版. - - 北京:北京邮电大学出版社,2015.10
ISBN 978-7-5635-4417-2

Ⅰ.①大… Ⅱ.①张… Ⅲ.①物理学—双语教学—高等学校—题解 Ⅳ.①O4-44

中国版本图书馆CIP数据核字(2015)第169119号

书　　　名:	A Guide to the Problems in University Physics(Second Edition)
著作责任者:	张晓光　芦鹏飞　席丽霞　韩利红　张　勇　编著
责 任 编 辑:	马晓仟
出 版 发 行:	北京邮电大学出版社
社　　　址:	北京市海淀区西土城路10号(邮编:100876)
发 行 部:	电话:010-62282185　传真:010-62283578
E-mail:	publish@bupt.edu.cn
经　　　销:	各地新华书店
印　　　刷:	北京源海印刷有限责任公司
开　　　本:	787 mm×1 092 mm　1/16
印　　　张:	13.75
字　　　数:	360 千字
版　　　次:	2010年9月第1版　2015年10月第2版　2015年10月第1次印刷

ISBN 978-7-5635-4417-2　　　　　　　　　　　　　　　定　价:30.00元
・如有印装质量问题,请与北京邮电大学出版社发行部联系・

第2版 前 言

《大学物理双语课程解题指导书》自 2010 年 9 月出版以来,已经历了 4 个多年头。我们已经感受到读者对本书的喜爱,在当当网上,读者对本书的评分是五星。读者对本书的评价是:"内容翔实的大物书!""全英文教程,用语准确,恰当。适合留学生出国准备使用。""如果是上双语的大学物理,那么这本书的知识点还是蛮全的,题量也比较大,可以用来很好地练习。"这是我们教授"大学物理双语课程"以来,第一次尝试写一本《大学物理双语课程解题指导书》。因此,我们将这几年在大学物理双语课程教学实践中的热情都倾注到了本书的写作上。读者对本书的喜爱是我们最大的回报。

此次再版,除了改正了一些文法错误、修正了一些物理概念的定义之外,第 15 章在结构上做了较大幅度的改动。我们认为对图表的合理运用,可以大大提高本书的可读性。因此本版增加了一些图表,修正了许多不完美的图,配合文字,能帮助读者更加容易地理解内容。在"Questions and Problems"部分,本版删除、增加或者替换了一些习题。

本版第 1~3 章由席丽霞编写,第 4~6 章由芦鹏飞编写,第 7~11 章由韩利红编写,第 12~14 章由张晓光编写,第 15~16 章由张勇编写,全书由张晓光统稿。

鉴于编者水平所限,本书中错误和不妥之处在所难免,恳请广大读者斧正。

编 者

2015 年 3 月

第 1 版 前 言

2001 年教育部在《关于加强高等学校本科教学工作提高教学质量的若干意见》中明确要求,高等学校的"本科教学要创造条件,引进原版外语教材,使用英语等外语进行公共课和专业课教学"。近年来,我国许多高校都开始试行公共课程与专业课程的双语教学。

物理学是一门自然科学,研究自然界物质的运动形式和运动规律,有一套完整的研究认识规律,与使用的语言无关,语言只是研究物理和描述自然的工具,不能替代科学的思维。但是由于社会和历史的原因,近几百年来科学技术的成果大多是用英语发表的。不论是科学技术专业杂志,还是各种国际学术会议,大都将英语作为交流语言。从这个意义上可以说"现代科学技术的研究主要是用英语思考",而现代科学技术大多是物理学的研究成果或者相关分支学科,因此"大学物理"课程采用双语教学十分重要,是有益的教学形式。

北京邮电大学早在 2005 年就开始了"大学物理"双语课的尝试,其背景是 2004 年北京邮电大学和英国伦敦大学玛丽女王学院联合创立学士学位联合培养项目,建立了国际学院,全部课程均用双语或者全英语教学。成绩合格的毕业生将分别获得北京邮电大学与英国伦敦大学玛丽女王学院的双学位。"大学物理"双语课是北京邮电大学国际学院的重要的基础课,由北京邮电大学理学院的资深教师承担教学任务,几年来积累了丰富的双语教学经验。

由于中西方的教学模式和理念均有很大的不同,因此国内外的"大学物理"课程的广度和深度不尽相同。为了与国内"大学物理"课程教学相适应,学生在掌握物理学的基本理论的基础之上,还必须通过一定题目的训练,从而达到巩固和提高的目的。由于英文教材的深度所致,题目的难度与中文教材的题目难度相比有不足,因此有必要编写一整套适合中国"大学物理"双语教学难度的,涵盖选择、填空和计算题等习题形式的学习指导书,从而有效地弥补现有教学的不足,成为学生学习"大学物理"双语课程中的一个有效组成部分。

编写本书就是为了达到上述目的所作的一次有益的尝试,指导接受"大学物理"双语课教学的学生如何用英语复习掌握所学内容,如何开阔视野,了解物理题的解题思路。全书总共安排了 16 章,涵盖了力学、电磁学、热学、振动与波动、波动光学、近代物理的内容。每一章的结构包括三部分内容。第一部分为知识点复习(Review of Contents),将相关物理概念、定律、定理作了较为详细的复习与梳理。第二部分为 Typical Examples,其中包括解题思路(Problem Solving Strategies)和典型例题(Examples)的详细解答。第三部分为 Questions and Problems,给出了一批有一定难度的题目,以选择、填空、问答以及计算题的形式给出。这样,学生在做练习的同时,也熟悉了各种考试采用英语的出题形式。在本书的最后给出了所有题目的参考答案。

本书第 1~3 章由席丽霞编写,第 4~6 章由芦鹏飞编写,第 7~11 章由韩利红编写,第 12~14 章由张晓光编写,第 15~16 章由金光生编写。全书由张晓光统稿。

由于编写《大学物理双语课程解题指导书》是一项新的尝试,也由于编者自身的水平有限,一定存在许多缺点与不足,欢迎读者批评指正。

编 者
2010 年 8 月

CONTENTS

Chapter 1　Describing Motion: Kinematics of Particles ·················· 1

　Review of the Contents ·················· 1
　Typical Examples ·················· 4
　　Problem solving strategy ·················· 4
　　Examples ·················· 5
　Questions and Problems ·················· 9

Chapter 2　Dynamics of Particles and Systems of Particles ·················· 13

　Review of the Contents ·················· 13
　Typical Examples ·················· 21
　　Problem solving strategy ·················· 21
　　Examples ·················· 22
　Questions and Problems ·················· 29

Chapter 3　Rotational Motion and Rigid Body ·················· 36

　Review of the Contents ·················· 36
　Typical Examples ·················· 39
　　Problem solving strategy ·················· 39
　　Examples ·················· 40
　Questions and Problems ·················· 45

Chapter 4　Electrostatics ·················· 52

　Review of the Contents ·················· 52
　Typical Examples ·················· 54
　　Problem solving strategy ·················· 54
　　Examples ·················· 55
　Questions and Problems ·················· 57

Chapter 5　Magnetism ·················· 63

　Review of the Contents ·················· 63
　Typical Examples ·················· 66

Problem solving strategy ··· 66
　　Examples ··· 66
　Questions and Problems ·· 68

Chapter 6　Electromagnetic Inductance ································· 74

　Review of the Contents ·· 74
　Typical Examples ··· 78
　　Problem solving strategy ··· 78
　　Examples ··· 78
　Questions and Problems ·· 80

Chapter 7　Kinetic Theory of Gases ··· 87

　Review of the Contents ·· 87
　Typical Examples ··· 89
　　Problem solving strategy ··· 89
　　Examples ··· 90
　Questions and Problems ·· 91

Chapter 8　The First Law of Thermodynamics ······················· 93

　Review of the Contents ·· 93
　Typical Examples ··· 97
　　Problem solving strategy ··· 97
　　Examples ··· 98
　Questions and Problems ·· 100

Chapter 9　The Second Law of Thermodynamics ······················· 103

　Review of the Contents ·· 103
　Typical Examples ··· 105
　　Problem solving strategy ··· 105
　　Examples ··· 106
　Questions and Problems ·· 108

Chapter 10　Oscillations ·· 110

　Review of the Contents ·· 110
　Typical Examples ··· 113
　　Problem solving strategy ··· 113

CONTENTS

    Examples ··· 113

    Questions and Problems ··· 114

Chapter 11   Wave Motion ··· 118

    Review of the Contents ··· 118

    Typical Examples ··· 123

        Problem solving strategy ··· 123

        Examples ··· 123

    Questions and Problems ··· 124

Chapter 12   Interference of Light ·· 127

    Review of the Contents ··· 127

    Typical Examples ··· 131

        Problem solving strategy ··· 131

        Examples ··· 131

    Questions and Problems ··· 134

Chapter 13   Diffraction of Light ·· 138

    Review of the Contents ··· 138

    Typical Examples ··· 145

        Problem solving strategy ··· 145

        Examples ··· 146

    Questions and Problems ··· 150

Chapter 14   Polarization of Light ·· 153

    Review of the Contents ··· 153

    Typical Examples ··· 160

        Problem solving strategy ··· 160

        Examples ··· 160

    Questions and Problems ··· 162

Chapter 15   Special Theory of Relativity ·· 165

    Review of the Contents ··· 165

    Typical Examples ··· 168

        Problem solving strategy ··· 168

        Examples ··· 169

Questions and Problems ·· 171

Chapter 16   Fundamentals of Quantum Theory ·· 174
  Review of the Contents ·· 174
  Typical Examples ·· 178
    Problem solving strategy ·· 178
    Examples ·· 179
  Questions and Problems ·· 182

Answers to Questions and Problems ·· 185

Chapter 1 Describing Motion: Kinematics of Particles

Review of the Contents

Mechanics or study of motion and its causes, is usually subdivided: **kinematics**, which is the description of how objects move, and **dynamics**, which deals with force and why objects move as they do. In present chapter, we start by discussing objects that move without rotating, such motion is called **translational motion**. That is to say, only translational motion is discussed. We often use a **particle** as an idealized model for a moving body when effects such as rotation or change of shape are not important. Particle is considered to be a mathematical point and to have no spatial extent (no size), which can undergo only translational motion. To describe the position of an object, the other object referred to should be chosen, which is called **reference frame**. It is arbitrary. In everyday life, we usually choose the earth as reference frame. We also need a **coordinate system** to describe the position of an object. Cartesian coordinate system and natural coordinate system are used in this chapter. In kinematics of particles, we describe the motion of a particle by using vectors to specify its position, velocity, and acceleration.

1. The position vector and displacement of a particle

Consider a particle that is at point P, **position vector** r of the particle at this instant is a vector that goes from the origin to the point P, as shown in Fig. 1-1. In Cartesian coordinate system, position vector can be written as

$$r = x\hat{i} + y\hat{j} + z\hat{k} \tag{1-1}$$

As the particle moves through space, position vector varies with time t, so $r = r(t)$ is called **motion function**. In Cartesian coordinate system, motion function can be written as

$$r = r(t) = x(t)\hat{i} + y(t)\hat{j} + z(t)\hat{k} \tag{1-2}$$

The path that a particle moves through space is in general a curve (Fig. 1-1), which is called **trajectory equation**, cancelling time factor, we get the trajectory equation

$$f(x, y, z) = 0 \tag{1-3}$$

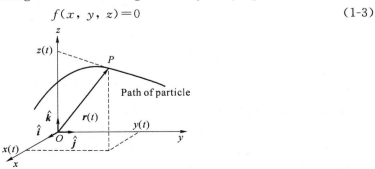

Fig. 1-1 The description of particle's position vector

During a time interval Δt the particle moves from P_1, where its position vector is r_1, to P_2, where its position vector is r_2 (shown in Fig. 1-2).

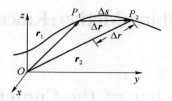

Fig. 1-2 The difference between displacement and distance

The change in position (called **displacement**) during this time interval is
$$\Delta r = r_2 - r_1 = r_2(t + \Delta t) - r_1(t) \tag{1-4}$$
For a Cartesian system, displacement can be written as
$$\Delta r = (x_2 - x_1)\hat{i} + (y_2 - y_1)\hat{j} + (z_2 - z_1)\hat{k} = \Delta x \hat{i} + \Delta y \hat{j} + \Delta z \hat{k} \tag{1-5}$$

Notes:

(1) Displacement is a **vector**, the magnitude of this vector is
$$|\Delta r| = \sqrt{(x_2 - x_1)^2 + (y_2 - y_1)^2 + (z_2 - z_1)^2}$$

(2) Displacement is independent on the choice of origin.

(3) Displacement is different from the distance. **Distance** Δs is the total lengths of the path curve (Fig. 1-2), which is scalar. Generally, $|\Delta r| \neq \Delta s$, but for infinitesimal, $|dr| = ds$.

(4) Displacement is different from the quantity Δr (Fig. 1-2). Δr is the difference between the magnitude of the position vectors r_2 and r_1.
$$\Delta r = |r_2| - |r_1| = r_2 - r_1 \neq |\Delta r| = |r_2 - r_1|$$

2. Velocity and speed

If a particle's displacement during a time interval Δt is Δr as shown in Fig. 1-3, the **average velocity** during this interval is defined to be:
$$\bar{v} = \frac{\Delta r}{\Delta t} \tag{1-6}$$

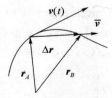

Fig. 1-3 Instantaneous velocity and average velocity

In Cartesian coordinate, average velocity is described as:
$$\bar{v} = \frac{\Delta x}{\Delta t}\hat{i} + \frac{\Delta y}{\Delta t}\hat{j} + \frac{\Delta z}{\Delta t}\hat{k} \tag{1-7}$$

The definition of the **instantaneous velocity** is the limit of the average velocity as Δt

Chapter 1 Describing Motion: Kinematics of Particles

approaches zero:

$$v = \lim_{\Delta t \to 0} \frac{\Delta \boldsymbol{r}}{\Delta t} = \frac{\mathrm{d}\boldsymbol{r}}{\mathrm{d}t} \tag{1-8}$$

The direction of $\boldsymbol{v}$ at any moment is along the line tangent to the path at that moment. In Cartesian coordinate, the component description of velocity $\boldsymbol{v}$ is:

$$\boldsymbol{v} = \frac{\mathrm{d}x}{\mathrm{d}t}\boldsymbol{i} + \frac{\mathrm{d}y}{\mathrm{d}t}\boldsymbol{j} + \frac{\mathrm{d}z}{\mathrm{d}t}\boldsymbol{k} = v_x \hat{\boldsymbol{i}} + v_y \hat{\boldsymbol{j}} + v_z \hat{\boldsymbol{k}} \tag{1-9}$$

The **speed** v is defined to be:

$$v = \lim_{\Delta t \to 0} \frac{\Delta s}{\Delta t} = \frac{\mathrm{d}s}{\mathrm{d}t} = \left|\frac{\mathrm{d}\boldsymbol{r}}{\mathrm{d}t}\right| = |\boldsymbol{v}| \tag{1-10}$$

The magnitude of the vector $\boldsymbol{v}$ at any instant is the speed v of the particle at that instant. In Cartesian coordinate system, speed v can be written as:

$$v = \sqrt{(v_x)^2 + (v_y)^2 + (v_z)^2} = \sqrt{\left(\frac{\mathrm{d}x}{\mathrm{d}t}\right)^2 + \left(\frac{\mathrm{d}y}{\mathrm{d}t}\right)^2 + \left(\frac{\mathrm{d}z}{\mathrm{d}t}\right)^2} \tag{1-11}$$

In natural coordinate system, velocity is:

$$\boldsymbol{v} = v\,\hat{\boldsymbol{\tau}} = \frac{\mathrm{d}s}{\mathrm{d}t}\hat{\boldsymbol{\tau}} \tag{1-12}$$

3. Acceleration

The **average acceleration** over a time interval Δt is defined as:

$$\bar{\boldsymbol{a}} = \frac{\Delta \boldsymbol{v}}{\Delta t} \tag{1-13}$$

The definition of **instantaneous acceleration** is the limit of the average acceleration as the time interval Δt is allowed to approach zero:

$$\boldsymbol{a} = \lim_{\Delta t \to 0} \frac{\Delta \boldsymbol{v}}{\Delta t} = \frac{\mathrm{d}\boldsymbol{v}}{\mathrm{d}t} = \frac{\mathrm{d}^2 \boldsymbol{r}}{\mathrm{d}t^2} \tag{1-14}$$

Acceleration in Cartesian coordinate system:

$$\boldsymbol{a} = a_x \hat{\boldsymbol{i}} + a_y \hat{\boldsymbol{j}} + a_z \hat{\boldsymbol{k}} = \frac{\mathrm{d}v_x}{\mathrm{d}t}\hat{\boldsymbol{i}} + \frac{\mathrm{d}v_y}{\mathrm{d}t}\hat{\boldsymbol{j}} + \frac{\mathrm{d}v_z}{\mathrm{d}t}\hat{\boldsymbol{k}} = \frac{\mathrm{d}^2 x}{\mathrm{d}t^2}\hat{\boldsymbol{i}} + \frac{\mathrm{d}^2 y}{\mathrm{d}t^2}\hat{\boldsymbol{j}} + \frac{\mathrm{d}^2 z}{\mathrm{d}t^2}\hat{\boldsymbol{k}} \tag{1-15}$$

4. Circular motion

A particle moves in a circular path of radius r with constant speed v is said to be in **uniform circular motion**, which has a **centripetal (or normal) acceleration** a_n toward the center of the circle.

For a uniform circular motion, in natural coordinate system, the velocity and acceleration are:

$$\boldsymbol{v} = v\,\hat{\boldsymbol{\tau}} = \frac{\mathrm{d}s}{\mathrm{d}t}\hat{\boldsymbol{\tau}} \quad \text{and} \quad \boldsymbol{a} = \boldsymbol{a}_\mathrm{n} = \frac{v^2}{r}\hat{\boldsymbol{n}} \tag{1-16}$$

If the speed of a particle revolving in a circle is changing, the motion is called **non-uniform circular motion**, as shown in Fig. 1-4. There will be a tangential acceleration as well as the normal acceleration. In natural coordinate, the total acceleration is:

A Guide to the Problems in University Physics (Second Edition)

$$a = a_n + a_t = \frac{v^2}{r}\hat{n} + \frac{dv}{dt}\hat{\tau} \tag{1-17}$$

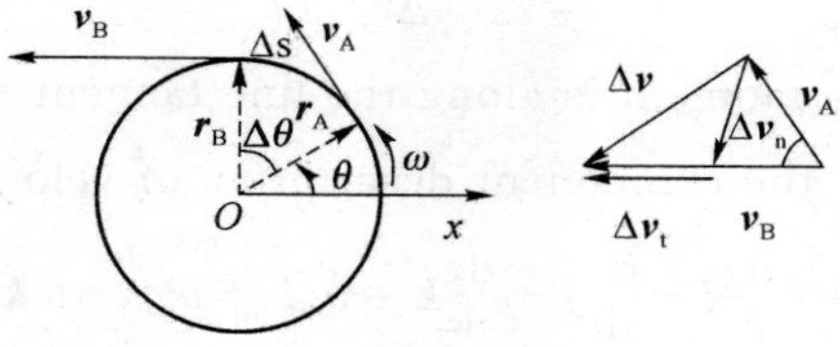

Fig. 1-4 The change in velocity Δv can be divided into Δv_n plus Δv_t

Normal acceleration $\quad a_n = \lim\limits_{\Delta t \to 0}\dfrac{\Delta v_n}{\Delta t} = \dfrac{v^2}{r}\hat{n}$

—due to the change in direction of the velocity vector

Tangential acceleration $a_t = \lim\limits_{\Delta t \to 0}\dfrac{\Delta v_t}{\Delta t} = \lim\limits_{\Delta t \to 0}\dfrac{\Delta |v|}{\Delta t} = \dfrac{dv}{dt}\hat{\tau}$

—arises from the change in magnitude of the velocity vector

5. The relative motion respect to two frames in translation

There are two reference frames S and S', as shown in Fig. 1-5, the position vector and velocity of O' relative to O are $r_{O'O}$ and $v_{O'O}$ respectively. Considering a moving particle P, the relationship between position vectors and velocities of P in these two reference frames can be described as:

$$r_{PO} = r_{PO'} + r_{O'O} \quad \text{and} \quad v_{PO} = v_{PO'} + v_{O'O} \tag{1-18}$$

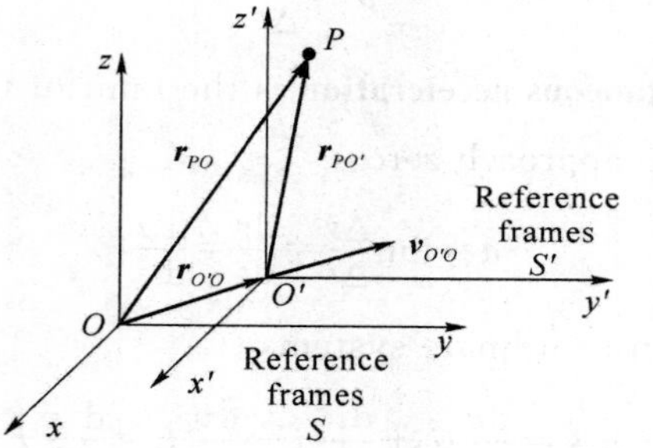

Fig. 1-5 The descriptions of position vectors of particle P relative to two reference frames

Typical Examples

Problem solving strategy

There are two categories of problems in Kinematics.

1. The motion function of particle is known, its velocity and acceleration at any moment are needed to find, we can solve the problems by way of derivatives, $v = \dfrac{dr}{dt}$ and $a = \dfrac{dv}{dt} = \dfrac{d^2 r}{dt^2}$.

2. The acceleration of particle and initial conditions such as the initial velocity and initial position vector ($v(0), r(0)$) are known quantities, its velocity and position vector at any

Chapter 1 Describing Motion: Kinematics of Particles

moment are needed to find, we can use the way of integrals, $\boldsymbol{v} = \boldsymbol{v}_0 + \int_{t_0}^{t} \boldsymbol{a} \, dt$ and $\boldsymbol{r} = \boldsymbol{r}_0 + \int_{t_0}^{t} \boldsymbol{v} \, dt$. In fact, we usually calculate velocity and position vector by components integrals.

$$v_x = v_{x0} + \int_{t_0}^{t} a_x \, dt, \quad v_y = v_{y0} + \int_{t_0}^{t} a_y \, dt, \quad v_z = v_{z0} + \int_{t_0}^{t} a_z \, dt$$

$$x = x_0 + \int_{t_0}^{t} v_x \, dt, \quad y = y_0 + \int_{t_0}^{t} v_y \, dt, \quad z = z_0 + \int_{t_0}^{t} v_z \, dt$$

Examples

1. A particle moves with the motional function as: $\boldsymbol{r} = 2t\,\hat{\boldsymbol{i}} + (2 - t^2)\,\hat{\boldsymbol{j}}$ (SI), as shown in Fig. 1-6.

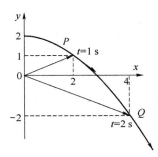

Fig. 1-6 Example 1

Find: (1) its trajectory function;

(2) its velocities at $t = 1$ s and 2 s respectively;

(3) its accelerations at $t = 1$ s and 2 s respectively;

(4) the path distance it travels during this time interval.

Solution:

(1) Here $x = 2t$ and $y = 2 - t^2$, canceling time t, getting the trajectory function $y = 2 - \frac{1}{4}x^2$ (parabola)

(2) According to the definition of velocity, we can get: $\boldsymbol{v} = \dfrac{d\boldsymbol{r}}{dt} = 2\,\hat{\boldsymbol{i}} - 2t\,\hat{\boldsymbol{j}}$.

at $t = 1$ s, velocity $\boldsymbol{v}_1 = 2\boldsymbol{i} - 2\boldsymbol{j}$;

at $t = 2$ s, velocity $\boldsymbol{v}_2 = 2\boldsymbol{i} - 4\boldsymbol{j}$.

(3) According to the definition of acceleration, we can get: $\boldsymbol{a} = -2\boldsymbol{j}$.

It is a constant acceleration, so at $t = 1$ s and 2 s, $\boldsymbol{a}_1 = \boldsymbol{a}_2 = \boldsymbol{a} = -2\boldsymbol{j}$.

(4) The path distance $\Delta s = \int_{P}^{Q} ds$

$$ds = \sqrt{(dx)^2 + (dy)^2} = \sqrt{(2dt)^2 + (-2t\,dt)^2} = 2\sqrt{1 + t^2}\,dt$$

So during time interval from $t = 1$ s to $t = 2$ s, the traveling path distance is:

$$\Delta s = \int_{P}^{Q} ds = \int_{1}^{2} 2\sqrt{1 + t^2}\,dt = 3.62 \text{ m}$$

Comparing with the magnitude of displacement:

$$r_1 = 2i + 1j, \quad r_2 = 4i - 2j, \quad |\Delta r| = \sqrt{2^2 + 3^2} = 3.61 \text{ m} < \Delta s$$

2. Find the velocity and position at any time of a particle in projectile motion, with initial velocity v_0 and initial position O, air resistance is neglected, as shown in Fig. 1-7.

Solution:

We choose a reference frame fixed with respect to the Earth, its origin O being the release point. The acceleration of particle and initial conditions are known, by way of integrals, velocity and position can be found.

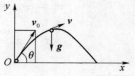

Fig. 1-7 Example 2

$$a = \frac{d\boldsymbol{v}}{dt} \quad \text{and} \quad \boldsymbol{a} = -g\boldsymbol{j}$$

$$\int_{v_0}^{v} d\boldsymbol{v} = \int_{t_0}^{t} \boldsymbol{a} dt = \int_{t_0}^{t} (-g\boldsymbol{j}) dt$$

$$\boldsymbol{v} = \boldsymbol{v}_0 - gt\boldsymbol{j}$$

$$\boldsymbol{r} = \int_0^r d\boldsymbol{r} = \int_0^t \boldsymbol{v} dt \qquad \boldsymbol{r} = \boldsymbol{v}_0 t - \frac{1}{2} g t^2$$

In Cartesian coordinate system:

$$v_{0x} = v_0 \cos\theta; \quad v_{0y} = v_0 \sin\theta; \quad a_x = 0; \quad a_y = -g$$

$$\boldsymbol{v} = (v_0 \cos\theta) \boldsymbol{i} + (v_0 \sin\theta - gt) \boldsymbol{j} = v_x \boldsymbol{i} + v_y \boldsymbol{j}$$

$$\boldsymbol{r} = (v_0 t \cos\theta) \boldsymbol{i} + \left(v_0 t \sin\theta - \frac{1}{2} g t^2\right) \boldsymbol{j}$$

Trajectory:
$$y = x \tan\theta - \frac{1}{2} \frac{g x^2}{v_0^2 \cos^2\theta}$$

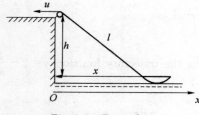

Fig. 1-8 Example 3

3. A person on a cliff pulls a boat floating in water with a constant speed u through a rope over a pulley fixed on the edge of the cliff. The height of cliff above water is h, and the horizontal distance between the cliff and the boat is x as shown in Fig. 1-8. Find the velocity and acceleration of the boat in water.

Solution:

Take right side to be positive.

Starting from the relation:

$$l^2 = h^2 + x^2 \tag{1-19}$$

Take the time derivatives of equation (1-19) two sides, we can get

$$2l \frac{dl}{dt} = 2x \frac{dx}{dt} \tag{1-20}$$

Notice:
$$\frac{dl}{dt} = -u \quad \frac{dx}{dt} = v \tag{1-21}$$

Substitute Equation (1-20) by (1-21), we get the velocity of the boat

$$v = -\frac{l}{x} u = -\frac{\sqrt{h^2 + x^2}}{x} u = -\frac{u}{\cos\theta}$$

Chapter 1 Describing Motion: Kinematics of Particles

(Hint: "—" means the moved direction of the boat is negative to x axis)

The acceleration of the boat is $a = \dfrac{dv}{dt} = \dfrac{dv}{dx}\dfrac{dx}{dt} = v\dfrac{dv}{dx} = -\dfrac{h^2}{x^3}u^2$.

4. A ladder of length l leans against a vertical wall. The bottom end of the ladder slides to the right with the constant speed of u. Find the velocities and accelerations of points A and M ($|MB|=b$) when $|OB|=X$.

Solution:

Establish Cartesian coordinate system, as shown in Fig. 1-9.

(1) For point A, only slides along y axis, which location is satisfied

$$X^2 + Y^2 = l^2 \tag{1-22}$$

Take the time derivatives of equation (1-22) two sides, we can get

$$2X\dfrac{dX}{dt} + 2Y\dfrac{dY}{dt} = 0 \tag{1-23}$$

So the velocity of point A is $\boldsymbol{v}_A = v_{Ay}\hat{\boldsymbol{j}} = \dfrac{dY}{dt}\hat{\boldsymbol{j}} = -\dfrac{X}{Y}u\hat{\boldsymbol{j}} = -\dfrac{X}{\sqrt{l^2-X^2}}u\hat{\boldsymbol{j}}$

The acceleration is $\boldsymbol{a}_A = a_{Ay}\hat{\boldsymbol{j}} = \dfrac{dv_{Ay}}{dt}\hat{\boldsymbol{j}} = \dfrac{dv_{Ay}}{dX}\dfrac{dX}{dt}\hat{\boldsymbol{j}} = -\dfrac{l^2 u^2}{(l^2-X^2)^{3/2}}\hat{\boldsymbol{j}}$.

(2) For point M, the location is satisfied: $\dfrac{X-x}{b} = \dfrac{X}{l}$ and $\dfrac{Y}{l} = \dfrac{y}{b}$.

The velocity is $v_{Mx} = \dfrac{l-b}{l}u$, and $v_{My} = \dfrac{b}{l}v_{Ay} = -\dfrac{bX}{l\sqrt{l^2-X^2}}u$

The acceleration $a_{Mx} = \dfrac{dv_{Mx}}{dt} = \dfrac{l-b}{l}\dfrac{du}{dt} = 0$, $a_{My} = \dfrac{dv_{My}}{dt} = \dfrac{b}{l}\dfrac{dv_{Ay}}{dt} = -\dfrac{blu^2}{(l^2-X^2)^{3/2}}$

5. A particle moves along a circle of radius of R as shown in Fig. 1-10. The path it follows is $s = v_0 t - \dfrac{1}{2}bt^2$, where v_0 and b are positive constant ($v_0^2 > Rb$). Find:

(1) When will $|a_t| = a_n$;

(2) When will the magnitude of acceleration equal b;

(3) How many revolutions that the particle has completed when magnitude of acceleration reaches to b?

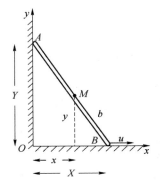

Fig. 1-9 Example 4

Fig. 1-10 Example 5

Solution:

(1) The path of the particle is known, we can get the speed of the particle at any time t, the speed is:
$$v = \frac{ds}{dt} = v_0 - bt$$

The normal acceleration is: $a_n = \dfrac{v^2}{R} = \dfrac{(v_0 - bt)^2}{R}$.

The tangential acceleration is: $a_t = \dfrac{dv}{dt} = -b$.

Let $|a_t| = a_n$, time t satisfy: $t = \dfrac{v_0}{b} \pm \sqrt{\dfrac{R}{b}}$.

(2) The magnitude of acceleration is $a = \sqrt{a_t^2 + a_n^2} = \sqrt{b^2 + \dfrac{(v_0 - bt)^4}{R^2}}$.

Let $a = b$, solve time $t = \dfrac{v_0}{b}$.

(3) When $t = \dfrac{v_0}{b}$, $s = v_0 \dfrac{v_0}{b} - \dfrac{1}{2} b \dfrac{v_0^2}{b^2} = \dfrac{v_0^2}{2b}$.

The particle completed the revolutions $N = \dfrac{s}{2\pi R} = \dfrac{v_0^2}{4\pi bR}$.

6. A wheel of radius R rolls on the ground without slipping, its center moves with a constant u. Find the motional equation of a point A on the rim of the wheel.

Solution:

Establish two reference frames, wheel and the ground as shown in Fig. 1-11.

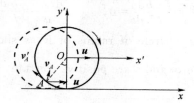

Fig. 1-11   Example 6

In the reference frame of wheel, the velocity of point A is:
$$v'_{Ax} = -\omega R \cos \omega t$$
$$v'_{Ay} = \omega R \sin \omega t$$

In the reference frame of ground, the velocity of point A satisfies
$$v_{Ax} = v'_{Ax} + u = -\omega R \cos \omega t + u$$
$$v_{Ay} = v'_{Ay} = \omega R \sin \omega t$$

At $t = 0$, $v_A(0) = 0$, so we get $u = \omega R$.

The velocity of point A can be written:
$$v_{Ax} = -u \cos \frac{u}{R} t + u$$
$$v_{Ay} = u \sin \frac{u}{R} t$$

By way of integrals, we get the motional equation of point A:

$$x = \int_0^R v_{ax}\,dt + x_0 = \int_0^t \left(-u\cos\frac{u}{R}t + u\right)dt = ut - R\sin\frac{u}{R}t$$

$$y = \int_0^R v_{ay}\,dt + y_0 = \int_0^t \left(u\sin\frac{u}{R}t\right)dt = R\left(1 - \cos\frac{u}{R}t\right)$$

The rim moves in the path of a cycloid.

Questions and Problems

1. An object is moving with velocity given by $v(t) = v_x(t)\hat{i} + v_y(t)\hat{j} + v_z(t)\hat{k}$, where $v_z(t) = 0$. From this, one can conclude

   (1) that the acceleration $a(t)$ _____.

   (A) will have no components that identically zero

   (B) may have some components that are identically zero

   (C) will have only a z component that is identically zero

   (D) will have an identically zero z component, and maybe an identically zero component in the x or y direction

   (2) and the position $r(t)$ _____.

   (A) will have no components that are identically zero

   (B) may have some components that are identically zero

   (C) will have only a z component that is identically zero

   (D) will have an identically zero z component, and maybe an identically zero component in the x or y direction

2. An object is moving in the xy plane with the position as a function of time given by $r = x(t)\hat{i} + y(t)\hat{j}$. Point O is at $r = 0$. The object is definitely moving toward O when _____.

   (A) $v_x > 0, v_y > 0$ (B) $v_x < 0, v_y < 0$ (C) $xv_x + yv_y < 0$ (D) $xv_x + yv_y > 0$

3. An object is launched straight up into the air from the ground with an initial vertical velocity of 30 m/s. The object rises to a highest point approximately 45 m above the ground in 3 seconds; it then falls back to the ground in 3 more seconds, impacting with a speed of 30 m/s.

   (1) The average speed of the object during the 6-second interval is closest to _____.

   (A) 0 m/s   (B) 5 m/s   (C) 15 m/s   (D) 30 m/s

   (2) The magnitude of the average velocity during the 6-second interval is closest to _____.

   (A) 0 m/s   (B) 5 m/s   (C) 15 m/s   (D) 30 m/s

4. An object is moving along the x axis with position as a function of time given by $x = x(t)$. Point O is at $x = 0$. The object is definitely moving toward O when _____.

   (A) $dx/dt < 0$ (B) $dx/dt > 0$ (C) $d(x^2)/dt < 0$ (D) $d(x^2)/dt > 0$

5. An object starts from rest at $x = 0$ when $t = 0$. The object moves in the x direction with positive velocity after $t = 0$. The instantaneous velocity and average velocity are related by _____.

   (A) $dx/dt < x/t$

(B) $dx/dt = x/t$

(C) $dx/dt > x/t$

(D) dx/dt can be larger than, smaller than, or equal to x/t

6. An object is moving in the x direction with velocity $v_x(t)$, and dv_x/dt is a nonzero constant. With $v_x = 0$ when $t = 0$, then for $t > 0$ the quantity $v_x dv_x/dt$ is _____.

(A) negative

(B) zero

(C) positive

(D) not determined from the information given

7. An object moves with a constant acceleration $\boldsymbol{a}$. Which of the following expressions are also constant (but not equal 0)? _____.

(A) $d|\boldsymbol{v}|/dt$ (B) $|d\boldsymbol{v}/dt|$ (C) $d(v^2)/dt$ (D) $d(\boldsymbol{v}/|\boldsymbol{v}|)/dt$

8. For a particle moved along an arbitrary curved path, which statement is most correct? _____.

(A) The tangential acceleration is not zero

(B) The normal acceleration is not zero

(C) Because the velocity only has tangential component, the normal acceleration must be zero

(D) If the particle moves with a constant speed, the magnitude of the total acceleration must be zero

9. An object moves along an arbitrary curved path, $\boldsymbol{r}$ represents position vector, s is distance, v is speed, a_t is tangential acceleration, for the following expressions, which statement is correct? _____.

(1) $\dfrac{dv}{dt} = a$ (2) $\dfrac{dr}{dt} = v$ (3) $\dfrac{ds}{dt} = v$ (4) $\left|\dfrac{d\boldsymbol{v}}{dt}\right| = a_t$

(A) (1) and (4) are correct   (B) (2) and (4) are correct

(C) only (2) is correct   (D) only (3) is correct

10. A particle moves along a circle of radius of R with a variable speed v, the magnitude of the acceleration for the particle is _____.

(A) $\dfrac{dv}{dt}$

(B) $\dfrac{v^2}{R}$

(C) $\dfrac{dv}{dt} + \dfrac{v^2}{R}$

(D) $\left[\left(\dfrac{dv}{dt}\right)^2 + \left(\dfrac{v^4}{R^2}\right)\right]^{1/2}$

11. A particle moves along the direction of x axis, $a = 2t$. At $t = 0, x_0 = 0, v_0 = 0$. At $t = 2$ s, its velocity is _____, the position vector is _____.

12. A particle moves along x direction and $a = 2 + 6x^2$. At $t = 0, x_0 = 0, v_0 = 0$. The velocity $v(x) = $ _____.

13. The position of a particle is given by $\boldsymbol{r} = (6.0\cos 3.0t \hat{\boldsymbol{i}} + 6.0\sin 3.0t \hat{\boldsymbol{j}})$ meters. Determine (1) the velocity vector $\boldsymbol{v} = $ _____, (2) the acceleration vector $\boldsymbol{a} = $ _____, (3) the path of this particle _____, (4) the relation between $\boldsymbol{r}$ and $\boldsymbol{a}$ is _____.

Chapter 1 Describing Motion: Kinematics of Particles

14. A stone is shot with speed of 30 m/s and in horizontal direction. At $t=5.0$ s, the magnitude of its tangential acceleration is _____, and the magnitude of its normal acceleration is _____.

15. A wheel of radius R rotates about horizontal axis O_1, a block is mounted on the fringe of the wheel by a string as shown in Fig. 1-12. If the motion function of the block is $y=\frac{1}{2}bt^2$, what are the $\mathbf{v}(t)$ and $\mathbf{a}(t)$ at any point M. (Suppose at $t=0$, M coincides with O'). Please using the natural frame to describe motion of M.

16. A particle starts from the origin at $t=0$ with an initial velocity of 5.0 m/s along the positive x axis, if the acceleration is $(-3.0\mathbf{i}+4.5\mathbf{j})$ m/s^2, determine the velocity and position of the particle at the moment it reaches its maximum x coordinate.

17. A particle moves in xy-plane. Its motional equations are $x(t)=R\cos \omega t$ and $y(t)=R\sin \omega t$, where R and ω are constant. Determine: (1) Show that the particle moves in a circle of radius R; (2) Show that the magnitude of the particle's velocity is constant and equals ωR; (3) Show that the particle's acceleration is always opposite to its position vector and has the magnitude of $\omega^2 R$.

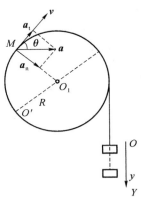

Fig. 1-12 Problem 15

18. A projectile is fired from point A with an initial velocity v_0 which forms an angle α with the horizontal as shown in Fig. 1-13. Find the radius of curvature of the trajectory of the projectile at point B and C.

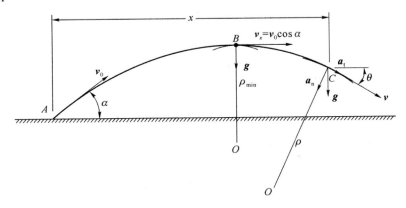

Fig. 1-13 Problem 18

19. A balloon moves up from ground with an initial vertical velocity of v_0. For the reason of wind, in the air the balloon is blew to the right with horizontal velocity $v_x=by$ (b is a positive constant, y is the height of the balloon). Choose the right side to be positive for x axis.

(1) Find the motional equation of the balloon.

(2) Find the path (trajectory) equation of balloon.

(3) Determine the tangential acceleration and the radius of the curvature of the trajectory with respect to height y.

20. Determine the velocity as a function of time for a body falling vertically from rest when there is a resistive force linearly proportional to v, such as $\dfrac{dv}{dt} = g - \dfrac{b}{m}v$.

21. A woman can row a boat at 4 km/h in still water, (1) If she is crossing a river where the current is 2 km/h, in what direction must her boat be headed if she wants to reach a point directly opposite her stating point (2) If the river is 4 km wide, how long will it take her to cross the river?

Chapter 2 Dynamics of Particles and Systems of Particles

Review of the Contents

In this chapter, we discuss the cause of motion, a field of study called **dynamics**. The principles of dynamics can be wrapped up in a neat package of three statements called **Newton's laws of motion**. Three theorems derived from Newton's laws and laws of conservation are also introduced.

1. Newton's first law

Content: When no force acts on a body, or when the vector sum of all forces acting on it (the net force) is zero, the body is in equilibrium. If the body is initially at rest, it remains at rest; if it is initially in motion, it continues to move with constant velocity.

From Newton's first law, we can get the following conclusions:

(1) An object has a tendency to maintain its original state of motion in the absence of a force. This tendency is called **inertia**. The inertial property of a body is characterized by its **mass**.

(2) The force is the only reason which makes the states of body change. It indicates the concept of **force**: An interaction can cause an acceleration of a body.

(3) Defining a special set of reference frames called **inertial frames**—in which Newton's first law (also second law) is valid. Any reference frame that moves with constant velocity with respect to an inertial frame is also an inertial frame. Reference frames where the law of inertia does not hold, such as the accelerating reference frames are called non-inertial frames. In most situations, we consider a reference frame connected to the Earth as the approximate inertial frame.

2. Newton's second law

Content: The acceleration of a body is directly proportional to the net force acting on it and is inversely proportional to its mass.

$$\sum \boldsymbol{F} = m\boldsymbol{a} \qquad (2\text{-}1)$$

Notes:

(1) An **instantaneous relation**: Once the force acting on a body changes (whether in magnitude or in direction), the acceleration also changes at that moment. Once the force acting on it vanished, the acceleration becomes zero immediately.

(2) It is only suitable for the inertial frame, and particles or particle-like bodies.

(3) The component expressions:

In Cartesian coordinate system

$$\sum F_x = ma_x \quad \sum F_y = ma_y \quad \sum F_z = ma_z \tag{2-2}$$

In natural coordinate system

$$\sum F_t = ma_t = m\frac{dv}{dt} \quad \sum F_n = ma_n = m\frac{v^2}{\rho} \tag{2-3}$$

3. Newton's third law

Content: If body 1 exerts a force on body 2 (an "action"), body 2 exerts a force on body 1 (a "reaction"). These two forces have the same magnitude but are opposite in direction. These two forces act on different bodies.

$$\boldsymbol{F}_{12} = -\boldsymbol{F}_{21} \tag{2-4}$$

We should note that, in this statement, "action" and "reaction" are the two opposite forces, sometimes refer to them as an **action-reaction pair**, this is not meant to imply any cause-and-effect relationship; the two forces are identical and simultaneous, never act on the same body and do not cancel each other.

4. Impulse and momentum

Newton's laws are useful for solving a wide range of problems in dynamics. However, there is one class of problems in which, even though Newton's law still applies, we may have insufficient knowledge of the forces to permit us to analyze the motion, these problems involve collisions between one object and another. In order to solve such problems, impulse, linear momentum and the law of conservation of linear momentum are used.

- **Impulse**

For any arbitrary force $\boldsymbol{F}$, the integral of the force over the time interval during which it acts is called the **impulse** $\boldsymbol{J}$:

$$\boldsymbol{J} = \int_{t_1}^{t_2} \boldsymbol{F} dt \tag{2-5}$$

For the definition of impulse, we should note that:

(1) The impulse depends on the strength of the force and on its duration of time.

(2) The impulse of a force is a vector. When the force is constant, the direction of $\boldsymbol{J}$ is as same as the force; if $\boldsymbol{F}$ is variable, the direction of $\boldsymbol{J}$ is determined by the integral of $\int_{t_1}^{t_2} \boldsymbol{F} dt$

(3) When a time-varying net force $\boldsymbol{F}$ is difficult to measure, we can use a time-averaged net force as the substitute provided that it would give the same impulse to the particle in the same time interval.

$$\bar{\boldsymbol{F}} = \frac{1}{\Delta t}\int_{t_i}^{t_f} \boldsymbol{F} dt \tag{2-6}$$

- **Momentum (or linear momentum)**

The momentum $\boldsymbol{p}$ of a particle is defined as the product of its mass and its velocity,

$$\boldsymbol{p} = m\boldsymbol{v} \tag{2-7}$$

- **Impulse-momentum theorem for a particle**

The impulse of the net force acting on a particle during a given time interval is equal to the change in momentum of the particle during that interval.

Chapter 2 Dynamics of Particles and Systems of Particles

$$J = \int_{t_1}^{t_2} F \, dt = p_2 - p_1 = \Delta p \tag{2-8}$$

Impulse-momentum theorem is only valid in inertial reference frame. In reality, we often use its component form (t_f means final time, t_i means initial time):

$$\left. \begin{array}{l} J_x = \int_{t_i}^{t_f} F_x \, dt = mv_{fx} - mv_{ix} \\[4pt] J_y = \int_{t_i}^{t_f} F_y \, dt = mv_{fy} - mv_{iy} \\[4pt] J_z = \int_{t_i}^{t_f} F_z \, dt = mv_{fz} - mv_{iz} \end{array} \right\} \tag{2-9}$$

- **Impulse-momentum theorem for a system of particles**

The total momentum of a system of particles is the vector sum of the momentum of the individual particles $p_{tot} = \sum_i p_i$. The total external force acting on the system is $\sum_i F_{i\,ext}$.

The derivative form:

$$\sum_i F_{i\,ext} = \frac{dp_{tot}}{dt} \tag{2-10}$$

The integral form:

$$\int_{t_1}^{t_2} \sum_i F_{i\,ext} \, dt = p_{tot\,2} - p_{tot\,1} \tag{2-11}$$

Here, we should note the role of the internal forces, which can exchange the momentum between particles within system, but can not influence the total momentum of the system.

- **Conservation of momentum**

When the vector sum of external forces on a system is zero, the total momentum of the system is constant.

$$\text{When } \sum_i F_{i\,ext} = 0, \; p_{tot} = \sum_i p_i = \text{constant vector} \tag{2-12}$$

When one of the components of sum of external forces is zero, the component of total momentum is conserved in this direction—conservation of momentum in component form.

$$\sum_i F_{i\,ext\text{-}x} = 0, \; p_{tot\text{-}x} = \sum_i p_{ix} = \text{constant} \tag{2-13}$$

- **Center of mass**

We can restate the principle of conservation of momentum in a useful way by using the concept of **center of mass**. For a system of particles (as shown in Fig. 2-1), with masses m_1, m_2, and so on, the position vector is r_i of the i-th particle, the position vector r_{CM} of the center of mass is defined as

$$r_{CM} = \frac{m_1 r_1 + m_2 r_2 + m_3 r_3 + \cdots}{m_1 + m_2 + m_3 + \cdots} = \frac{\sum_i m_i r_i}{\sum_i m_i} \tag{2-14}$$

In coordinate system, the components description is:

$$x_{CM} = \frac{\sum_i m_i x_i}{M} \quad y_{CM} = \frac{\sum_i m_i y_i}{M} \quad z_{CM} = \frac{\sum_i m_i z_i}{M} \tag{2-15}$$

For the extended object with uniformly distribution of mass (as shown in Fig. 2-2), the position vector r_{CM} of the center of mass can be written as

$$r_{CM} = \frac{1}{M}\int r \, dm \tag{2-16}$$

In Cartesian coordinate system,

$$x_{CM} = \frac{1}{M}\int x \, dm, \quad y_{CM} = \frac{1}{M}\int y \, dm, \quad z_{CM} = \frac{1}{M}\int z \, dm \tag{2-17}$$

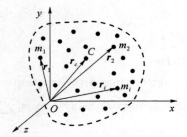
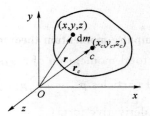

Fig. 2-1 The center of mass for a system of particles

Fig. 2-2 The center of mass for an extended object

- **Motion of the center of mass**

The velocity of the center of mass is

$$v_{CM} = \frac{m_1 v_1 + m_2 v_2 + m_3 v_3 + \cdots}{m_1 + m_2 + m_3 + \cdots} = \frac{\sum_i m_i v_i}{\sum_i m_i} \tag{2-18}$$

The total momentum of the system of particles is equal to its total mass times the velocity of center of mass, just as though all the mass were concentrated at center of mass, that is

$$p_{tot} = \sum_i m_i v_i = M v_{CM} \tag{2-19}$$

If the net external force on the system is not zero, the center of mass accelerates and satisfies

$$\sum F_{ext} = M a_{CM} \tag{2-20}$$

—Newton's second law for the motion of center of mass

If the net external force on the system is zero, the center of mass velocity v_{CM} is constant and the total momentum of the system is conserved.

5. Work and energy

We discuss an alternative analysis of the motion of an object in terms of the quantity energy. Firstly, considering one particular form—kinetic energy, the energy associated with a body because of its motion, and then introducing the concept of work, which is related to kinetic energy through the work-energy theorem. This theorem, derived from Newton's laws, provides new and different insight into the behavior of mechanical systems. At last, we introduce another kind of energy—potential energy, and develop a conservation law for energy.

- **Kinetic energy**

The kinetic energy of a particle with mass m and speed v is $K = \frac{1}{2}mv^2$.

Chapter 2　Dynamics of Particles and Systems of Particles

The kinetic energy of a system of particles is the sum of each particle's kinetic energy,
$$K = \sum_i K_i = \sum_i \frac{1}{2} m_i v_i^2 \tag{2-21}$$

- **Work**

Work done by a constant force along a straight-line displacement is
$$W = F|\Delta r|\cos\theta = \mathbf{F} \cdot \Delta \mathbf{r} \tag{2-22}$$

Work done by a varying force along a curve path

As shown in Fig. 2-3, divide the path into a large number of small displacement. In each segment of displacement the work can be considered as done by a constant force along a straight-line displacement $dW = \mathbf{F} \cdot d\mathbf{r}$, the total work done by the force is
$$W = \int_A^B \mathbf{F} \cdot d\mathbf{r} \tag{2-23}$$

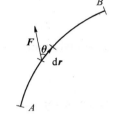

Fig. 2-3　The calculation of work

About the definition of the work, we should note that:
(1) Work is a scalar quantity, no direction.
(2) Work is a process quantity. Generally, depends on the path followed by the particle. Different path corresponds to different work done by the same force.
(3) Calculation of work relates to the reference frame.
(4) If there are many forces acting on a particle, the total work done by multiple forces is the scalar addition of the work done by each force.
$$W_{net} = \int_A^B \mathbf{F}_{net} \cdot d\mathbf{l} = \int_A^B \left(\sum_i \mathbf{F}_i\right) \cdot d\mathbf{l} = \sum_i \int_A^B \mathbf{F}_i \cdot d\mathbf{l} = \sum_i W_i \tag{2-24}$$

(5) Component description

In Cartesian coordinate system,
$$W = \int_A^B \mathbf{F} \cdot d\mathbf{r} = \int_A^B (F_x dx + F_y dy + F_z dz) = \int_{x_A}^{x_B} F_x dx + \int_{y_A}^{y_B} F_y dy + \int_{z_A}^{z_B} F_z dz \tag{2-25}$$

In natural coordinate system,
$$W = \int_A^B (\mathbf{F}_n + \mathbf{F}_t) \cdot ds\hat{\tau} = \int_A^B F_t ds \tag{2-26}$$

The work done by a pair of internal forces $f_{12} = -f_{21}$

For an infinitesimal process shown in Fig. 2-4.
$$dW = \mathbf{f}_{12} \cdot d\mathbf{r}_1 + \mathbf{f}_{21} \cdot d\mathbf{r}_2 = \mathbf{f}_{21} \cdot (d\mathbf{r}_2 - d\mathbf{r}_1) = \mathbf{f}_{21} \cdot d(\mathbf{r}_2 - \mathbf{r}_1) = \mathbf{f}_{21} \cdot d\mathbf{r}_{21} \tag{2-27}$$

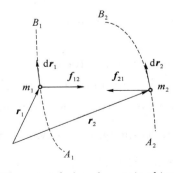

Fig. 2-4　The network done by a pair of internal forces

The calculation of net work done by a pair of internal forces on two particles is equivalent to—in the reference frame of particle 1, the calculation of work done by one force acting on particle 2, which **is independent of reference frame.**

Power

Power is the time rate of doing work. If an amount of work ΔW is done in a time interval Δt, the **average power** P_{av} is

$$P_{av} = \frac{\Delta W}{\Delta t} \tag{2-28}$$

The **instantaneous power** is defined as

$$P = \lim_{\Delta t \to 0} \frac{\Delta W}{\Delta t} = \frac{dW}{dt} \tag{2-29}$$

When a force F acts on a particle moving with velocity v, the instantaneous power is

$$P = \boldsymbol{F} \cdot \boldsymbol{v} \tag{2-30}$$

Work-kinetic energy theorem

Work-kinetic energy theorem of a particle

The work done by the net force on a particle equals the change in kinetic energy

$$W_{net} = K_2 - K_1 = \Delta K \tag{2-31}$$

Here, we should note the relationship between work and kinetic energy.

(1) Work is the measurements for the change in the kinetic energy "K" of a body. If $W_{net} > 0$, then $K_2 > K_1$, body gains the kinetic energy; If $W_{net} < 0$, then $K_2 < K_1$, body loses "K" and does work outside.

(2) Both work and "K" are scalars. They have same units and dimensions; "K" only depends on the speed of initial and final states, but work depends on the real process.

(3) Only valid in the inertial frame of reference.

Work-kinetic energy theorem for a system of particles

The sum work done on a system by all external and internal forces equal to the change in kinetic energy of the system,

$$\sum W_{i\text{-ext}} + \sum W_{i\text{-int}} = \Delta K \tag{2-32}$$

Note:

Generally, the works done by internal forces between particles cannot be canceled, because the displacements of particles are different.

- **Potential energy**

The conservative force and non-conservative force

If the work done by a force is independent of the path and depends only on the initial and final positions, the force is **conservative force.** The equivalent statement is the total work done by a conservative force is zero as the particle moves along a round trip, $\oint \boldsymbol{F} \cdot d\boldsymbol{r} = 0$.

If the work done by a force is not the same for all paths which are from the same initial position to the same final position, we call the force **a non-conservative force.**

Chapter 2 Dynamics of Particles and Systems of Particles

Potential energy

The potential energy U is energy associated with the configuration of a system. If the work is done in a system by a conservative force, the configuration of its parts and the potential energy change. We define the change in potential energy associated with a conservative force as

$$\Delta U = U(\boldsymbol{r}_b) - U(\boldsymbol{r}_a) = -W = -\int_a^b \boldsymbol{F} \cdot \mathrm{d}\boldsymbol{r} \qquad (2\text{-}33)$$

The definition of potential energy only gives the change in potential energy, or the relative value of potential energy. We can choose a position $\boldsymbol{r}_0 = \boldsymbol{r}_a$ as the reference point, define $U(\boldsymbol{r}_0)=0$, potential energy can be written as

$$U(\boldsymbol{r}) = U(\boldsymbol{r}) - 0 = -\int_{\boldsymbol{r}_0}^{\boldsymbol{r}} \boldsymbol{F} \cdot \mathrm{d}\boldsymbol{r} \qquad (2\text{-}34)$$

In order to understand the potential energy, we should note that:

(1) The potential energy U is the energy associated with the configuration of a system. Here "configuration" means how the parts of a system are located or arranged with respect to one another (the compression or stretching of the spring in the block-spring system, or height of the ball in the ball-Earth system.)

(2) The potential energy belongs to the system. We should properly speak of "the elastic potential energy of the block-spring system" or "the gravitational potential energy of the ball-Earth system", not "the elastic potential energy of the spring" or "the gravitational energy of the ball".

Several potential energy

(1) For **gravitational potential energy** near the Earth's surface, it is accustomed to choosing the reference point $y_0 = 0$ as surface of the Earth,

$$U(y) = mgy \qquad (2\text{-}35)$$

(2) For gravitational potential energy associate with two particles, it is accustomed to taking $U(r_0 = \infty) = 0$,

$$U(r) = -G\frac{Mm}{r} \qquad (2\text{-}36)$$

(3) For elastic potential energy, it is accustomed to choosing the reference position to be that in which the spring is in its relaxed state,

$$U(x) = \frac{1}{2}kx^2 \qquad (2\text{-}37)$$

Force and potential energy

If a potential energy expression is given, the corresponding force is the negative of the gradient of the potential energy function.

$$\boldsymbol{F} = -\left(\hat{\boldsymbol{i}}\,\frac{\partial U}{\partial x} + \hat{\boldsymbol{j}}\,\frac{\partial U}{\partial y} + \hat{\boldsymbol{k}}\,\frac{\partial U}{\partial z}\right) = -\nabla U \qquad (2\text{-}38)$$

- **Work-energy theorem**

Mechanical energy: $E_{\text{mech}} = K + U$ is defined to be total mechanical energy of the system.

Work-energy theorem

The work done by all the external forces and internal non-conservative forces acting on a system of particles equals the change in total mechanical energy of the system.

$$\sum W_{i\text{-ext}} + \sum W_{i\text{-int-nonconserv}} = \Delta E_{\text{mech}} = E_{\text{mech 2}} - E_{\text{mech 1}} \tag{2-39}$$

- **Conservation of mechanical energy**

For a system, if $\sum W_{i\text{-ext}} + \sum W_{i\text{-int-nonconserv}} = 0$, then

$$\Delta E_{\text{mech}} = 0 \quad \text{or} \quad K + U = \text{constant} \tag{2-40}$$

Note:

(1) For a system in which only internal conservative forces act, the total mechanical energy remains constant.

(2) The internal conservative forces acting within the system only change kinetic energy into potential energy or potential energy into kinetic energy, do not influence the total mechanical energy.

6. Torque-Angular momentum theorem

The most general motion of a body includes rotational as well as translational motions. Here, we consider the causes of rotation by inducing the concepts of angular momentum, toque and Torque-Angular momentum theorem.

- **Angular momentum**

The angular momentum L, with respect to the point O, of a particle with mass m, velocity v, is defined as

$$\boldsymbol{L} = \boldsymbol{r} \times m\boldsymbol{v} \tag{2-41}$$

where r is the particle's position vector with respect to point O.

Here, we should note that, the angular momentum is a vector and depends on the choice of the reference point.

- **Torque**

Definition $\qquad\qquad\qquad \boldsymbol{M} = \boldsymbol{r} \times \boldsymbol{F} \tag{2-42}$

Torque is also a vector and depends on the choice of the reference point.

- **Torque-Angular momentum theorem for a particle**

The torque acting on a particle is equal to the time rate of change of the particle's angular momentum.

$$\boldsymbol{M} = \frac{d\boldsymbol{L}}{dt} \tag{2-43}$$

Here, the origins of M and L must be same and the reference frame must be inertial frame.

- **Torque-Angular momentum theorem for a system of particles**

The net external torque acting on the system is equal to the time rate of change of the total angular momentum of the system.

$$\sum \boldsymbol{M}_{\text{ext}} = \sum_i \frac{d\boldsymbol{L}_i}{dt} = \frac{d}{dt}\sum_i \boldsymbol{L}_i = \frac{d\boldsymbol{L}_{\text{tot}}}{dt} \tag{2-44}$$

We should note that the torques of each pair of internal forces are vanished each other,

Chapter 2 Dynamics of Particles and Systems of Particles

so they don't change the total angular momentum of the system. All the origins of $\boldsymbol{M}$ and $\boldsymbol{L}$ in the system are same.

- **Conservation of angular momentum**

For a system of particles, if the net external torque acting on the system is zero, the total angular momentum of a system remains constant. That is to say

if $$\sum \boldsymbol{M}_{\text{ext}} = 0, \text{ then } \boldsymbol{L}_{\text{tot}} = \text{constant} \tag{2-45}$$

Typical Examples

Problem solving strategy

(1) Isolate the object whose motion is being analyzed. Draw a separate free-body diagram for each object.
- Be sure to include all the forces acting on the object, but be equally careful not to include any force exerted by this object on other object.
- Never include the quantity $m\boldsymbol{a}$ in your free-body diagram. It's not a force.
- Identify the internal forces and external forces.

(2) Establish a convenient reference frame and an appropriate coordinate system attached to it.

(3) Judge whether the conditions of conservation of mechanical energy or momentum are satisfied. If $\sum W_{i\,\text{ext}} + \sum W_{i\,\text{int-nonconserv}} = 0$, the mechanical energy is conserved. When $\sum_i \boldsymbol{F}_{i\,\text{ext}} = 0$, the total momentum of the system is a constant vector, or one of the components of sum of external forces is zero, the component of total momentum is conserved in this direction—conservation of momentum in component form.

(4) When using the laws of conservation, decide what the initial and final states (the positions and velocities) of the system are. Then write the equations.
- Keep in mind, that the work done by internal conservative forces must be represented in the change in potential energy, dot not include them again in work done by force.

(5) If the conditions of conservation are not satisfied, considering using work-energy theorem or impulse-momentum theorem. Especially when the problem involves motion with varying forces, motion along a curved path or both. Choose the initial and final positions of the body, list the equations.

(6) Otherwise, for each object, write the equations for Newton's second law in component manner.
- Generally, the number of unknowns must be equal to the number of equations.
- If the number of unknowns $<$ the number of equations, there must be equivalent equations.
- If the number of unknowns $>$ the number of equations, find relationship between motions(this situation mostly occurs to many objects whose motions are dependent).

(7) Solve the equations to find unknowns.

(8) Check the result.

Examples

1. A small ball of mass m is attached to the end of a cord of length R, which rotates under the influence of gravitational force in vertical circle about a fixed point O.

(1) Determine the tension in the cord at any angle θ.

(2) When the ball starts motion in the bottom of the circle, in order to pass point B which is the top of the circle, find the minimum value of initial velocity v_0.

Solution:

The forces diagram for the small ball is shown in Fig. 2-5(a). Choose the Earth as reference frame. Establish natural coordinate system. List Newton's second law.

Tangential:
$$-mg\sin\theta = m\frac{dv}{dt} \tag{2-46}$$

Normal:
$$T - mg\cos\theta = m\frac{v^2}{R} \tag{2-47}$$

Unknown quantities: θ, T, v. Additional equation to be found according to circular motion,

$$v = R\omega = R\frac{d\theta}{dt} \tag{2-48}$$

Change the independent variable t to θ,

$$-mg\sin\theta = m\frac{dv}{d\theta}\frac{d\theta}{dt} = m\omega\frac{dv}{d\theta} = m\frac{v}{R}\frac{dv}{d\theta}$$

We get

$$\int_{v_0}^{v} mv\,dv = \int_{0}^{\theta} -mgR\sin\theta\,d\theta$$

$$\frac{1}{2}mv^2 = \frac{1}{2}mv_0^2 + mgR\cos\theta - mgR$$

The above equation can be obtained by using the theorem of conservation of mechanical energy.

$$T = \frac{mv_0^2 - mgR(2 - 3\cos\theta)}{R}$$

At the point B, $T \geqslant 0$, and $\theta = 180°$, that is

$$\frac{mv_0^2 - mgR(2+3)}{R} \geqslant 0, \quad v_0 \geqslant \sqrt{5gR}$$

Check the result: in the case of $\theta = 90°$, $\theta = 270°$, as shown in Fig. 2-5(b), the tangential acceleration will be $-g$ and g.

$$\frac{1}{2}mv^2 = \frac{1}{2}mv_0^2 - mgR + mgR\cos\theta$$

$$mv\frac{dv}{dt} = -mgR\sin\theta\frac{d\theta}{dt} = -mgR\omega\sin\theta$$

$$\frac{dv}{dt} = -g\sin\theta$$

$$\theta = 90°\quad a_t = -g$$
$$\theta = 270°\quad a_t = g$$

The results are expected.

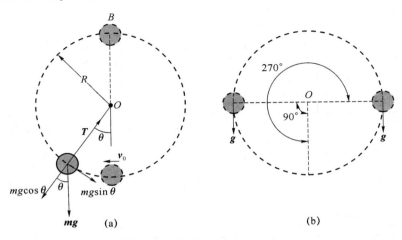

Fig. 2-5 Example 1

2. A block of mass m is put on a wedge M, which, in turn, is put on a horizontal table. The incline angle of wedge is θ. All surfaces are frictionless. Determine the accelerations of the block m and wedge M.

Solution:

Drawing the forces diagram for the block and wedge separately, as shown in Fig. 2-6(a). Choose the Earth as reference frame. Establish Cartesian coordinate system. List Newton's second law.

For m:

Horizontal $\qquad\qquad\qquad N_1 \sin\theta = ma_x$ (2-49)

Vertical $\qquad\qquad\qquad N_1 \cos\theta - mg = ma_y$ (2-50)

For M:

Horizontal $\qquad\qquad\qquad -N_1 \sin\theta = Ma_0$ (2-51)

Vertical $\qquad\qquad\qquad N_2 - N_1 \cos\theta - Mg = 0$ (2-52)

Unknown quantities: N_1, N_2, a_x, a_y, a_0

Motion relation (E means Earth): $\boldsymbol{a}_{mE} = \boldsymbol{a}_{mM} + \boldsymbol{a}_{ME}$

Horizontal $\qquad\qquad\qquad a_x = a'\cos\theta - a_0$ (2-53)

Vertical $\qquad\qquad\qquad a_y = -a'\sin\theta$ (2-54)

Combining equations (2-49), (2-50), (2-51), (2-52), (2-53) and (2-54), we can get

$$a_x = \frac{g\sin\theta\cos\theta}{1+\frac{m}{M}\sin^2\theta} \qquad a_y = \frac{-\left(1+\frac{m}{M}\right)g\sin^2\theta}{1+\frac{m}{M}\sin^2\theta} \qquad a_0 = -\frac{\frac{m}{M}g\sin\theta\cos\theta}{1+\frac{m}{M}\sin^2\theta}$$

Check the results:

(1) By dimensional analysis—reasonable.

(2) Directions are correct.

(3) By introducing extreme cases.

If $m \gg M$, $a_x \to 0$ $a_y \to -g$

If $M \gg m$, $a_0 \to 0$ $a_x \to g\sin\theta\cos\theta$ $a_y \to -g\sin^2\theta$ (as shown in Fig. 2-6(b)).

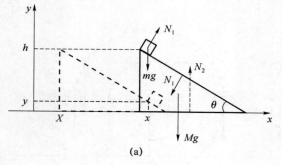 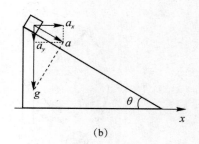

(a)   (b)

Fig. 2-6 Example 2

The results are reasonable.

3. A device called a capstan is used aboard ships in order to control a rope which is under great tension. The rope is wrapped around a fixed drum, usually for several turns. The load on the rope (end B) pulls it with a force T_B, and the sailor (end A) holds it with a much smaller force T_A. Show that $T_A = T_B \exp(-\theta\mu)$, where μ is the coefficient of friction and θ is the total angle subtended by the rope on the drum.

Solution:

Isolate an element of the rope to consider, the force diagram was shown in Fig. 2-7. Establish natural coordinate system. List Newton's second law.

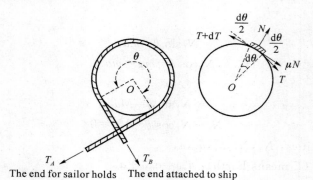

T_A T_B
The end for sailor holds   The end attached to ship

Fig. 2-7 Example 3

Tangential: $(T+dT)\cos(d\theta/2) - T\cos(d\theta/2) - \mu N = 0$ (2-55)

Normal: $(T+dT)\sin(d\theta/2) + T\sin(d\theta/2) - N = 0$ (2-56)

Here, for an element, it is satisfied with $\sin(d\theta/2) \approx d\theta/2$, $\cos(d\theta/2) \approx 1$, so (2-55) and (2-56) can be written as

$$dT - \mu N = 0$$

$$Td\theta + \frac{1}{2}dT d\theta - N = 0$$

Chapter 2 Dynamics of Particles and Systems of Particles

Neglect the second order infinitesimal $dTd\theta$, we get

$$Td\theta = \frac{dT}{\mu} \quad \int_{T_A}^{T_B}\frac{dT}{T} = \int_0^\theta \mu d\theta$$

$$T_A = T_B \exp(-\mu\theta)$$

As long as the θ is large enough, we can get $T_A \ll T_B$.

4. Variable force F is maintained tangent to a frictionless semicircular surface. By a slowly varying force F, a block with mass of m is moved, and spring to which it is attached is stretched from position 1 to position 2 as shown in Fig. 2-8. The spring has negligible mass and force constant k. The end of the spring moves in an arc of radius a. Calculate the work done by the force F.

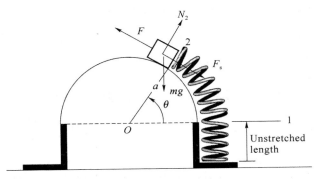

Fig. 2-8 Example 4

Solution 1: by integration directly

The block is in equilibrium in tangential direction:

$$F = ks + mg\cos\theta$$

$$W_F = \int_1^2 \mathbf{F}\cdot d\mathbf{s} = \int_0^s (ks + mg\cos\theta)ds$$

$$= \int_0^{a\theta} ks\,ds + \int_0^\theta mga\cos\theta\,d\theta = \frac{1}{2}ka^2\theta^2 + mga\sin\theta$$

Solution 2: by using work-energy theorem

External force: F; Internal forces: N(non-conservative, does no work), mg and F_s (conservative).

Choose the reference point at position 1 both for gravitational and elastic energy of block-spring-Earth system.

$$W_F = \Delta E = \Delta U = mga\sin\theta + \frac{1}{2}ks^2 = mga\sin\theta + \frac{1}{2}ka^2\theta^2$$

5. A ring of mass M hangs from a thread, and two beads of mass m slide on it without friction. The beads are released simultaneously from the top of the ring and slide down opposite sides. Show that the ring will start to rise if $m > 3M/2$, and find the angle at which this occurs.

Solution: Considering the beads and ring respectively.

For beads, see Fig. 2-9.

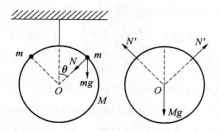

Fig. 2-9　Example 5

Normal component:
$$mg\cos\theta + N = m\frac{v^2}{R} \tag{2-57}$$

Conservation of mechanical energy: $mgR(1-\cos\theta) = \frac{1}{2}mv^2$ (2-58)

For ring, the condition under which it will rise: $2N'\cos\theta = 2N\cos\theta \geqslant Mg$ (2-59)

Unknowns: v, N, θ

Canceling v and N, we get

$$6m\cos^2\theta - 4m\cos\theta + M \leqslant 0$$

$$\frac{1}{3} - \frac{1}{3}\sqrt{1 - \frac{3}{2}\frac{M}{m}} \leqslant \cos\theta \leqslant \frac{1}{3} + \frac{1}{3}\sqrt{1 - \frac{3}{2}\frac{M}{m}} \tag{2-60}$$

Conclusion:

(1) After the ring starts to rise, the beads are no longer in circular motion. So the value of θ is chosen to be minimum.

$$\cos\theta = \frac{1}{3} + \frac{1}{3}\sqrt{1 - \frac{3}{2}\frac{M}{m}}$$

(2) In order for $\cos\theta$ be real, it must be $1 - \frac{3}{2}\frac{M}{m} > 0$, namely: $m > \frac{3}{2}M$.

6. Find the escape velocity v_{esc} of a satellite (the minimum initial speed needed to prevent it from returning to the Earth) of mass m, is projected into the air from the Earth, M_E is the mass of the Earth, R_E is the radius of the Earth.

Solution: Take the Earth and satellite as a system, there only exist conservative internal forces, so the mechanical energy of the system is conservative.

For initial state,
$$E = \frac{1}{2}mv_i^2 - \frac{GM_E m}{R_E} \tag{2-61}$$

For any state in the process $E = \frac{1}{2}mv^2 - \frac{GM_E m}{R_E + h} = \frac{1}{2}mv_i^2 - \frac{GM_E m}{R_E}$ (2-62)

And for any state, speed satisfies $v^2 = \frac{GM_E}{R_E + h}$ (2-63)

Considering $g = G\frac{M_E}{R_E^2}$, we get $v_i = \sqrt{v^2 - 2\frac{GM_E}{R_E + h} + 2\frac{GM_E}{R_E}}$

$$= \sqrt{gR_E\left(2 - \frac{R_E}{R_E + h}\right)} \tag{2-64}$$

From (2-64), we can see that: near the Earth, the greater the initial speed is, the higher the satellite can reach.

Chapter 2 Dynamics of Particles and Systems of Particles

When $R_E \gg h$, $v_i = v_{around} = \sqrt{gR_E} = 7.9 \times 10^3$ m/s, corresponding to the minimum initial speed of rotating around Earth.

If the speed is high enough, it will continue out into space never to return to Earth, then $h \to \infty$, with merely zero speed, the initial speed is corresponding to escape speed,

$$v_i v_{esc} = \sqrt{2GM_E/R_E} = \sqrt{2gR_E} = 11.2 \times 10^3 \text{ m/s}$$

7. A block of mass M slide along a horizontal table with speed v_0. At $x=0$ it hits a spring with spring constant k and begins to experience a friction force. The coefficient of friction is variable and is given by $\mu = bx$, where b is a constant. Find the loss in mechanical energy when the block has first come momentarily to rest.

Solution: Take block-spring-Earth and horizontal table as a system. No external force exists. Internal conservative forces: spring force, gravitational force, internal non-conservative forces: normal force (does no work), friction force.

Suppose the block's position is x_f at the moment when it first comes to rest. The work done by the friction force is:

$$W_{f_s}(x = 0 \to x_f) = \int_0^{x_f} -bMgx \, dx = -\frac{1}{2}bMgx_f^2$$

Using work-energy theorem: $-\frac{1}{2}bMgx_f^2 = \frac{1}{2}kx_f^2 - \frac{1}{2}Mv_0^2$

We get
$$x_f^2 = \frac{Mv_0^2}{k+bMg}$$

So
$$E_{loss} = E_i - E_f = -W_{f_s} = \frac{1}{2}bMgx_f^2 = \frac{bgM^2 v_0^2}{2(k+bMg)}$$

8. A chain of mass M length l is suspended vertically with its lowest end touching a scale. The chain is released and falls onto the scale as shown in Fig. 2-10. What is the reading of the scale when a length of chain, s, has fallen (Neglect the size of individual links)?

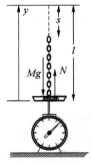

Fig. 2-10 Example 8

Solution:

(1) Using impulse-momentum theorem

Assuming a length of chain s has been already in the scale. Take an infinitesimal process during dt, a segment chain of length of ds impacts with the scale, and comes to a halt.

The impulse that the surface of the scale acting on this segment is:

$$F dt = 0 - (-dmv) = v\frac{M}{l} ds$$

$$F = \frac{M}{l}v\frac{ds}{dt} = \frac{M}{l}v^2 = \frac{M}{l}(2gs) = 2Mg\frac{s}{l}$$

The reading of the scale = the weight that has already in the scale $+ F = Mg\frac{s}{l} + 2Mg\frac{s}{l} = 3Mg\frac{s}{l}$

(2) Using the center of mass

Two part:
$$M_1 = \lambda(l-s) \quad y_1 = (l-s)/2$$
$$M_2 = \lambda s \quad y_2 = 0$$
$$\lambda = M/l$$
$$y_{CM} = \frac{M_1 y_1 + M_2 y_2}{M} = \frac{\lambda(l-s)(l-s)/2}{\lambda l} = \frac{(l-s)^2}{2l}$$

For the part in the scale: $s = \frac{1}{2}gt^2$, and the part in the air: $v = \frac{ds}{dt} = gt$

$$\frac{dy_{CM}}{dt} = \frac{d}{dt}\left[\frac{(l-s)^2}{2l}\right] = -\frac{(l-s)}{l}\frac{ds}{dt} = -\frac{gt}{l}\left(l - \frac{1}{2}gt^2\right)$$

$$\frac{d^2 y_{CM}}{dt^2} = \frac{g(3s-l)}{l}$$

Newton's second law for CM:

$$N - Mg = M\frac{d^2 y_{CM}}{dt^2} = Mg\left(\frac{3s}{l} - 1\right)$$

So
$$N = 3Mg\frac{s}{l}$$

9. A wooden block of mass M_1 is suspended from a cord of length L attached to a cart of mass M_2 which can roll freely on a frictionless horizontal track. A bullet of mass m is fired into the block from left. After the impact of the bullet, the block swings up with the maximum angle of θ, as shown in Fig. 2-11. What is the initial speed v of the bullet?

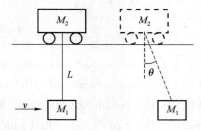

Fig. 2-11   Example 9

Solution:

Stage 1: For the system consisting of m and M_1, the momentum is conserved in horizontal during a small interval time of impact.

$$mv = (M_1 + m)v_1 \qquad (2-65)$$

Stage 2: The block plus bullet swing up with initial speed v_1, and drive the cart sliding forward in the track. At the moment when the block-bullet swing at maximum angle, $(M_1 + m)$ and M_2 have the same horizontal speed of v_2, and the mechanical energy of the system of $(M_1 + m)$ and M_2 is conserved.

$$\frac{1}{2}(M_1+m)v_1^2 = \frac{1}{2}(M_1+M_2+m)v_2^2 + (M_1+m)gL(1-\cos\theta) \qquad (2-66)$$

During the whole Stage1+Stage2: The momentum of system consisting of M_1, m, M_2 is conserved in horizontal.

$$mv = (M_1 + M_2 + m)v_2 \qquad (2-67)$$

Combine the equations (2-65), (2-66) and (2-67), we get the final answer,

Chapter 2 Dynamics of Particles and Systems of Particles

$$v = \frac{M_1+m}{m}\sqrt{\frac{M_1+M_2+m}{M_2}2gL(1-\cos\theta)}$$

10. Two boys, with same mass of m, suspend to the two side of a pulley with a light rope. The boy on the left makes an effort to climb up, but the other boy keeps at rest without any action as shown in Fig. 2-12. Which boy is the first to approach pulley? Neglect the mass of the pulley and the friction on the axis of the pulley.

Solution: Take the direction of torque consistent with anti-clockwise. For the two-boy system, the net external torque is zero.

$$\sum M_{\text{ext}} = Rm_1g - Rm_2g = 0$$

The angular momentum of two-boy system is conserved.

$$L_f = mR(v_2 - v_1) = L_i = 0$$

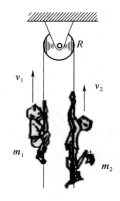

Fig. 2-12 Example 10

Two boys approach the pulley at the same time, whoever makes an effort.
But if $m_1 > m_2$,

$$\sum M_{\text{ext}} > 0, \frac{dL}{dt} > 0, L_i = 0, L_f > 0, v_2 > v_1$$

Questions and Problems

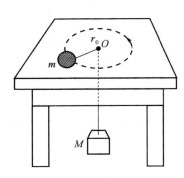

Fig. 2-13 Problem 1

1. A puck with mass m is moving in a circle of radius r_0 with a constant speed v_0 on a level frictionless table. A string is attached to the puck, which holds it in the circle; the string passes through a frictionless hole and is attached on the other end to a hanging object of mass M, as shown in Fig. 2-13.

(1) The puck is now made to move with a speed $v' = 2v_0$, but still in a circle. The mass of the hanging object is unchanged. The acceleration a' of the puck and the radius r' of the circle are now given by _____.

(A) $a' = 4a_0$ and $r' = r_0$ (B) $a' = 2a_0$ and $r' = r_0$
(C) $a' = 2a_0$ and $r' = 2r_0$ (D) $a' = a_0$ and $r' = 4r_0$

(2) The puck continues to move at speed $v' = 2v_0$ in a circle, but now the mass of the hanging object is doubled. The acceleration a' of the puck and the radius r' of the circle are now given by _____.

(A) $a' = 4a_0$ and $r' = r_0$ (B) $a' = 2a_0$ and $r' = r_0$
(C) $a' = 2a_0$ and $r' = 2r_0$ (D) $a' = a_0$ and $r' = 4r_0$

2. An object is moving in a circle at constant speed v. The magnitude of the rate of change of momentum of the object _____.

(A) is zero (B) is proportional to v
(C) is proportional to v^2 (D) is proportional to v^3

3. If the net force acting on a body is constant, what can we conclude about its momentum? _____.

(A) The magnitude and/or the direction of p may change
(B) The magnitude of p remains fixed, but its direction may change
(C) The direction of p remains fixed, but its magnitude may change
(D) p remains fixed in both magnitude and direction

4. An object is moving in a circle at constant speed v. From time t_i to time t_f, the object moves halfway around the circle. The magnitude of the impulse due to the net force on the object during this time interval is _____.

(A) zero (B) proportional to v
(C) proportional to v^2 (D) proportional to v^3

5. A variable force acts on object from 0 to t_f. The impulse of the force is zero. One can conclude that _____.

(A) $\Delta r=0$ and $\Delta p=0$ (B) $\Delta r=0$ but possibly $\Delta p=0$
(C) possibly $\Delta r \neq 0$ but $\Delta p=0$ (D) possibly $\Delta r \neq 0$ and possibly $\Delta p \neq 0$

6. Two objects are sitting on a level frictionless surface. The objects are not connected or touching. A force F is applied to one of the objects, which then moves with acceleration a. Which of the following statements is most correct? _____.

(A) The center of mass concept cannot be applied because the external force does not act on both objects
(B) The center of mass moves with acceleration that could be greater than a
(C) The center of mass moves with acceleration that must be equal to a
(D) The center of mass moves with acceleration that must be less than a

7. A system of N particles is free from any external forces.

(1) Which of the following is true for the magnitude of the total momentum of the system? _____.

(A) It must be zero
(B) It could be non-zero, but it must be constant
(C) It could be non-zero, and it might not be constant
(D) The answer depends on the nature of the internal forces in the system

(2) Which of the following must be true for the sum of the magnitudes of the momenta of the individual particles in the system? _____.

(A) It must be zero
(B) It could be non-zero, but it must be constant
(C) It could be non-zero, and it might not be constant
(D) It could be zero, even if the magnitude of the total momentum is not zero

8. A particle moves with constant momentum $p=(10 \text{ kg} \cdot \text{m/s})\hat{i}$. The particle has an angular momentum about the origin of $I=(20 \text{ kg} \cdot \text{m}^2/\text{s})\hat{k}$ when $t=0$ s.

Chapter 2 Dynamics of Particles and Systems of Particles

(1) The magnitude of the angular momentum of this particle is _____.

(A) decreasing   (B) constant

(C) increasing   (D) possibly but not necessarily constant

(2) The trajectory of this particle _____.

(A) definitely passes through the origin

(B) might pass through the origin

(C) will not pass through the origin, but uncertain how close it will pass to the origin

(D) will not pass through the origin, but one can calculate exactly how close it will pass to the origin

9. Two independent particles originally moving angular momenta L_1 and L_2 in a region of space with no external torques. A constant external torque M then acts on particle one, but not on particle two, for a time Δt. What is in the change in the total angular momentum of two particles? _____.

(A) $\Delta L = L_1 - L_2$

(B) $\Delta L = \frac{1}{2}(L_1 + L_2)$

(C) $\Delta L = M \Delta t$

(D) ΔL for the system is poorly defined because two particles are not connected

10. The force exerted by a special compression device is given by $F_x(x) = kx(x-l)$ for $0 \leqslant x \leqslant l$, where l is the maximum possible compression and k is a constant.

(1) The force required to compress the device a distant d is a maximum when _____.

(A) $d=0$ (B) $d=l/4$ (C) $d=l/\sqrt{2}$

(D) $d=l/2$ (E) $d=l$

(2) The work required to compress the device a distant d is a maximum when _____.

(A) $d=0$ (B) $d=l/4$ (C) $d=l/\sqrt{2}$

(D) $d=l/2$ (E) $d=l$

11. A 0.20 kg puck slides across a frictionless floor with a speed of 10 m/s. The puck strikes a wall and bounces off with a speed of 10 m/s in the opposite direction.

(1) The magnitude of the impulse on the puck is _____.

(A) 0 kg·m/s   (B) 1 kg·m/s   (C) 2 kg·m/s   (D) 4 kg·m/s

(2) The net work done on the puck is _____.

(A) -20 J   (B) -10 J   (C) 0 J   (D) 20 J

12. A block with mass m_1 is placed on an inclined plane with slope angel α and is connected to a second hanging block that has mass m_2 by a cord passing over a small, frictionless pulley (as shown in Fig. 2-14). The coefficient of static friction is μ_s, and the coefficient of kinetic friction is μ_k. (1) The mass m_2 is _____ for which block m_1 moves up the plane

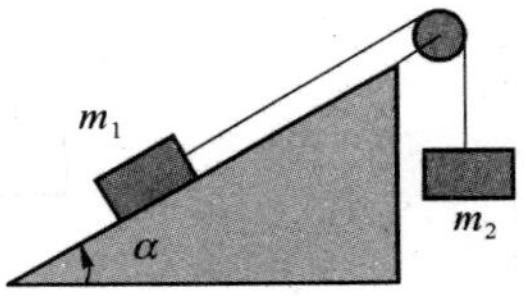

Fig. 2-14 Problem 12

at constant speed once it has been set in motion. (2) The mass m_2 is _____ for which block m_1 moves down the plane at constant speed once it has been set in motion. (3) The range of values of m_2 will be _____ when the blocks remain at rest if they are released from rest?

13. A child applies a force F parallel to the x-axis to a 10.0 kg sled moving on the frozen surface of a small pond. As the child controls the speed of the sled, the x-component of the force she applies varies with the x-coordinate of the sled as shown in Fig. 2-15. Suppose the sled is initially at rest at $x=0$. (1) When the sled moves from $x=0$ to $x=8.0$ m, the work done by the force F is _____, the speed of the sled is _____ at $x=8.0$ m; (2) When the sled moves from $x=8.0$ m to $x=12.0$ m, the work done by the force F is _____, the speed of the sled is _____ at $x=12.0$ m; (3) When the sled moves from $x=0$ to $x=12.0$ m, the work done by the force F is _____.

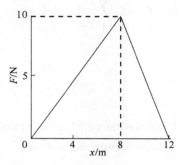

Fig. 2-15   Problem 13

14. A 4.00 kg block of ice is placed against a horizontal spring that has force constant $k=200$ N/m and is compressed 0.025 m. The spring is released and accelerates the block along a horizontal surface. Friction and the mass of the spring can be neglected. (1) During the motion of the block from its initial position to where the spring has returned to its uncompressed length, the work done on the block by the spring is _____. (2) The speed of the block is _____ after it leaves the spring.

15. Consider the system shown in Fig. 2-16. The rope and pulley have negligible mass, and the pulley is frictionless. Initially the 6.00 kg block is moving downward and the 8.00 kg block is moving to the right, both with a speed of 0.900 m/s. The blocks come to rest after moving 2.00 m. The coefficient of kinetic friction between the 8.00 kg block and the table top is _____.

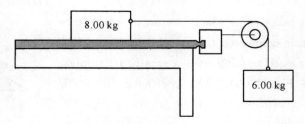

Fig. 2-16   Problem 15

Chapter 2 Dynamics of Particles and Systems of Particles

16. You throw a 0.145 kg baseball straight up in the air. Suppose your hand moves up 0.50 m while you are throwing the ball, which leaves your hand with an upward velocity of 20.0 m/s. Ignore air resistance. (1) Assuming that your hand exerts a constant upward force on the ball, the magnitude of that force is _____. (2) The speed of the ball is _____ at a point 15.0 m above the point where it leaves your hand.

17. A 12 kg crate sits on the floor. We want to load it into a truck by sliding it up a ramp 2.5 m long, inclined at 30°. A worker, giving no thought to friction, calculates that he can get the crate up the ramp by giving it an initial speed of 5.0 m/s at the bottom and letting it go. But friction is not negligible; the crate slides 1.6 m up the ramp, stops, and slides back down, as shown in Fig. 2-17. (1) Assuming that the friction force acting on the crate is constant, its magnitude is _____. (2) The speed of the crate is _____ when it reaches the bottom of the ramp.

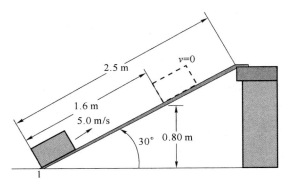

Fig. 2-17 Problem 17

18. A force parallel to the x-axis acts on a particle moving along the x-axis. This force produces a potential energy $U(x)$ given by $U(x) = \alpha x^4$, where $\alpha = 1.20$ J/m^4. When the particle is at $x = -0.800$ m, the force (magnitude and direction) is _____.

19. Coach Johnson's bat exerts a horizontal force on a 0.145 kg baseball of $\boldsymbol{F} = [(1.60 \times 10^7 \text{ N/s})t - (6.00 \times 10^9 \text{ N/s}^2)t^2]\hat{\boldsymbol{i}}$ between $t = 0$ and $t = 2.50$ ms. At $t = 0$ the baseball's velocity is $-(40.0\hat{\boldsymbol{i}} + 5.0\hat{\boldsymbol{j}})$ m/s. (1) The impulse exerted by the bat on the ball during the 2.50 ms that they are in contact is _____. (2) The impulse exerted by gravity on the ball during this time interval is _____. (3) The average force exerted by the bat on the ball during this time interval is _____. (4) The momentum is _____ and the velocity of the baseball is _____ at $t = 2.50$ ms.

20. A system consists of two particles. At $t = 0$ one particle is at the origin; the other, which has a mass of 0.50 kg, is on the y-axis at $y = 6.0$ m. At $t = 0$ the center of mass of the system is on the y-axis at $y = 2.4$ m. The velocity of the center of mass is given by $(0.75 \text{ m/s}^3)t^2 \hat{\boldsymbol{i}}$. (1) The total mass of the system is _____. (2) The acceleration of the center of mass at any time t is _____. (3) The net external force acting on the system is _____ at $t = 3.0$ s.

21. A chain has length L and mass M, and was put on a frictionless table. At $t=0$, the chain is stationary and length l is hanged over the edge. (1) What is the velocity at the moment that the whole chain leaves the table? (2) The time covered the whole process?

22. A force which acts on a particle moving in the xy plane is given by $\boldsymbol{F}=2y\,\hat{\boldsymbol{i}}+x^2\,\hat{\boldsymbol{j}}$ (SI). The particle moves from the origin to a final position C (5.00 m, 5.00 m). Calculate the work done by $\boldsymbol{F}$ along (1) OAC, (2) OBC, (3) OC, as shown in Fig. 2-18.

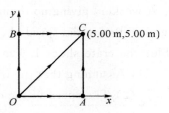

Fig. 2-18 Problem 22

23. A commonly used potential function to describe the interaction between the two atoms in a diatomic molecule is the Lennard-Jones potential, $U(x)=\varepsilon\left[\left(\dfrac{x_0}{x}\right)^{12}-2\left(\dfrac{x_0}{x}\right)^{6}\right]$. Find (1) the equilibrium separation between the atoms, (2) the force between the atoms, (3) the minimum energy necessary to break the molecule apart.

24. A small object of mass m is suspended from a string of length L. The object revolves in a horizontal circle of radium r with constant speed v. Determine the impulse exerted (1) by gravity; (2) by string tension on the object, during the time in which the object has passed half of the circle.

25. A small cube of mass m slides down a circular path of radius R cut into a large block of mass M. M rests on a table. M and m are initially at rest. m starts from the top of the path. Find the distance traveled by M when the cube m leaves the block M. (Suppose all the surfaces are frictionless.)

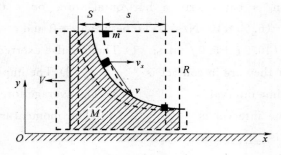

Fig. 2-19 Problem 25

26. A rocket is fired into the air. At the moment it reaches its highest point, a horizontal distance d from its starting point, an explosion separates it into two parts of equal mass. Part I is stopped in midair by explosion and falls vertically to Earth, as shown in Fig. 2-20. Where does part II land?

Chapter 2 Dynamics of Particles and Systems of Particles

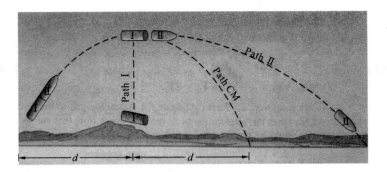

Fig. 2-20 Problem 26

27. A ball of mass m on a horizontal, frictionless table is connected to a string that passes through a small hole in the table. The ball is set into circular motion of radius R, at which time its speed is v_i. If the string is pulled from the bottom so that the radius of the circular path is decreased to r, what is final speed v_f of the ball?

Chapter 3   Rotational Motion and Rigid Body

Review of the Contents

In this chapter, we mainly deal with **rotational motion** of rigid bodies. A **rigid body** is an idealized model, which means a body with a perfectly definite and unchanging shape and size. The motion of a rigid body can be analyzed as the translational motion of its center of mass (CM) plus rotational motion about its center of mass (As shown in Fig. 3-1, a wheel rolling along the ground without slipping can be considered as translation of the wheel as a whole with velocity v_{CM} plus rotation about the CM). Translational motions have been discussed in detail in the previous chapter, so now we focus our attention on purely rotational motion. We begin with describing rotational motion. Then we will develop dynamic principles that relate the forces on a body to its rotational motion.

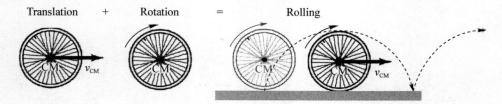

Fig. 3-1   A wheel rolling without slipping can be considered as translation of the wheel as a whole with velocity v_{CM} plus rotation about the CM

1. Description of a rigid body rotating about a fixed axis

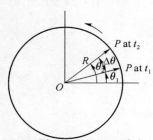

Fig. 3-2   The description of the rotational motion for a rigid body

When a rigid body rotates about a fixed axis, its position can be described by an **angular coordinate** θ, as shown in Fig. 3-2. The **angular velocity** ω is defined as the time derivative of the angular coordinate θ:

$$\omega = \lim_{\Delta t \to 0} \frac{\Delta \theta}{\Delta t} = \frac{d\theta}{dt} \tag{3-1}$$

The **angular acceleration** α is defined as the derivative of the angular velocity ω, or the second time derivative of the angular coordinate θ:

$$\alpha = \lim_{\Delta t \to 0} \frac{\Delta \omega}{\Delta t} = \frac{d\omega}{dt} = \frac{d^2 \theta}{dt^2} \tag{3-2}$$

Relating linear and angular kinematics

When a rigid body rotates about a fixed axis, every particle in the body moves in a

Chapter 3　Rotational Motion and Rigid Body

circular path. The circle lies in a plane perpendicular to the axis and is centered on axis. For any point P moves in a circle of radius R, at any time the angle θ and the arc length S are related by $S = R\theta$, relation between linear and angular speed is $v = R\omega$, tangential acceleration of point P is $a_t = R\alpha$, centripetal (normal) acceleration is $a_n = \omega^2 R$.

2. The moment of inertia I

The moment of inertia I of a body about a given axis is defined as:

$$I = m_1 r_1^2 + m_2 r_2^2 + \cdots = \sum_i m_i r_i^2 \tag{3-3}$$

where r_i is the perpendicular distance of mass m_i from the axis of rotation.

For continuous distribution bodies

$$I = \int r^2 \, dm \tag{3-4}$$

The moment of inertia reflects the tendency of a rigid body to resist angular acceleration, just like the mass reflecting the tendency of an object to resist linear acceleration. The greater the moment of inertia is, the more difficult it is to change the state of the body's rotation. An object's moment of inertia depends not only on the object's mass but on how the mass is distributed around the axis. Table 3-1 lists some typical rigid bodies' moment of inertia

Table 3-1　Moments of inertia of homogeneous rigid bodies with different geometries

Hoop or thin cylindrical shell $I_{CM} = mR^2$		Hollow cylinder $I_{CM} = \dfrac{1}{2} m(R_1^2 + R_2^2)$	
Solid cylinder or disk $I_{CM} = \dfrac{1}{2} mR^2$		Rectangular plate $I_{CM} = \dfrac{1}{12} m(a^2 + b^2)$	
Long thin rod with rotation axis through center $I_{CM} = \dfrac{1}{12} mL^2$		Long thin rod with rotation axis through end $I_{CM} = \dfrac{1}{3} mL^2$	

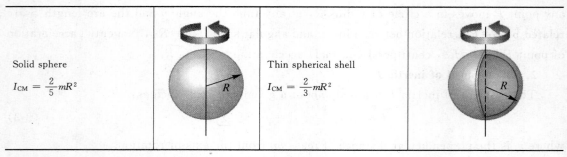

| Solid sphere $I_{CM} = \frac{2}{5}mR^2$ | Thin spherical shell $I_{CM} = \frac{2}{3}mR^2$ | |

The Parallel-axis theorem

The moment of inertia I_{CM} of a body of mass m about an axis through the center of mass is related to the moment of inertia I_p about a parallel axis at a distance d from the first axis by

$$I_p = I_{CM} + md^2 \tag{3-5}$$

The minimum of moment of inertia occurs at the axis through the center of mass as shown in Fig 3-3.

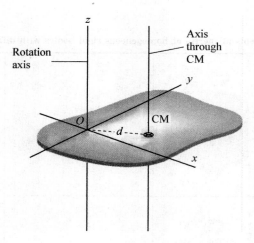

Fig. 3-3 The parallel-axis theorem

3. Newton's Second Law for rotation

The angular acceleration α of a rigid body rotating about a stationary axis is related to the net torque $\sum M$ acting on it and the moment of inertia I of the body, which satisfies

$$\sum M = I\alpha \tag{3-6}$$

In this law, the net external torque and the moment of inertia must be corresponding to the same axis.

We can see that Eq. (3-6) is analogous to Newton's Second Law, $\boldsymbol{F} = m\boldsymbol{a}$, for linear motion. Generally, this equation is valid for the rotation of a rigid body about a fixed axis in an inertial reference frame. But for the rotation about the axis fixed passing through the center of mass of the body, although the CM is not an inertial reference frame, it is also

Chapter 3   Rotational Motion and Rigid Body

valid. That is $\sum M_{\text{ext-CM}} = I_{\text{CM}}\alpha$.

4. Angular momentum for a rigid body about a fixed axis

- **Angular momentum** for a rigid body about a fixed axis can be written as

$$L_\omega = I\omega \tag{3-7}$$

where I is the moment of inertia about the same rotation axis.

- **The torque-angular momentum theorem**

The change rate of angular momentum for a rigid body about a fixed axis is equal to the net toque about that fixed axis acted on the rigid body. That is

$$\sum M_{\text{ext-axis}} = \frac{dL_\omega}{dt} \tag{3-8}$$

- **The conservation of angular momentum for rigid body**

When $\quad \sum M_{\text{ext-axis}} = 0, \ I\omega = I_0 \omega_0 \tag{3-9}$

The total angular momentum of rotating body remains constant if the net external torque acting on it is zero.

5. Work-kinetic energy theorem for a body rotating about a fixed axis

- **The kinetic energy** of a rigid body rotating about a fixed axis (rotational kinetic energy) is

$$K = \frac{1}{2} I \omega^2 \tag{3-10}$$

where I is the moment of inertia for that rotation axis.

- For a body with total mass m, the **gravitational potential energy** U is

$$U = mg y_{\text{CM}} \tag{3-11}$$

- An object that rotates while its center of mass undergoes translational motion will have both translational and rotational kinetic energy, which can be written as

$$K = K_{\text{cm}} + K_{\text{rot}} = \frac{1}{2} m v_{\text{cm}}^2 + \frac{1}{2} I_{\text{cm}} \omega^2 \tag{3-12}$$

- For a fixed axis rotation of a rigid body, the work done by a force can appear in the form of torque—**work done by a torque**.

$$W = \int_1^2 \boldsymbol{F} \cdot d\boldsymbol{l} = \int_1^2 F_{\tan} R d\theta = \int_{\theta_1}^{\theta_2} M d\theta \tag{3-13}$$

- **Work-kinetic energy theorem**

$$W_{\text{net}} = \int_{\theta_1}^{\theta_2} M_{\text{net}} d\theta = \int_{\omega_1}^{\omega_2} I\omega d\omega = \frac{1}{2} I \omega_2^2 - \frac{1}{2} I \omega_1^2 \tag{3-14}$$

The work done in rotating a body through an angle $\theta_2 - \theta_1$ is equal to the change in rotational kinetic energy of the body.

Typical Examples

Problem solving strategy

1. Draw a sketch of the situation, and select the body or bodies to be analyzed.

2. For each body, draw a free-body diagram isolating the body and including all the forces (and only those) that act on the body, including its weight. Label unknown quantities with algebraic symbols. A new consideration is that you must show the shape of the body accurately, including all dimensions and angles you will need for torque calculations.

3. Choose coordinate axes for each body, and indicate a positive sense of rotation for each rotating body. If there is a linear acceleration, it is usually simplest to pick a positive axis in its direction. If you know the sense of α in advance, picking that as the positive sense of rotation simplifies the calculations.

4. Judge whether the conditions of conservation of mechanical energy or angular momentum are satisfied. If $\sum W_{i\,\text{ext}} = 0$, the mechanical energy is conserved. When $\sum_i M_{i\text{-ext}} = 0$, the total angular momentum of the system is constant vector.

5. If the conditions of conservation are not satisfied, depending on the behavior of the body in question, apply $\sum F = ma$ or $\sum M = I\alpha$ or both.

6. If more than one body is involved, carry out Step 2 through 5 for each body. Write a separate equation of motion for each body. There may also be geometrical relations between the motions of two or more bodies.

7. Check the results for special cases or investigating extreme values of quantities when possible, and compare with your intuitive expectations.

Examples

1. Calculate the moment of inertia of a uniform hollow cylinder of inner radius R_1, outer radius R_2, and mass m, if the rotation axis is through the center along the axis of symmetry.

Solution: Divide the cylinder into thin concentric cylindrical rings or hoops of thickness dR, as shown in Fig. 3-4, then

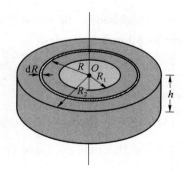

Fig. 3-4 Example 1

$$dm = \rho dV = \frac{m}{\pi(R_2^2 - R_1^2)h} 2\pi R h \, dR = \frac{2m}{R_2^2 - R_1^2} R \, dR$$

$$I = \int R^2 \, dm = \frac{2m}{R_2^2 - R_1^2} \int_{R_1}^{R_2} R^3 \, dR = \frac{2m}{R_2^2 - R_1^2} \cdot \frac{R_2^4 - R_1^4}{4} = \frac{1}{2} m (R_1^2 + R_2^2)$$

2. Uniform thin rod with mass m and length l. Calculate the moment of inertia about

Chapter 3   Rotational Motion and Rigid Body

the axis located (1) at the CM, (2) at an arbitrary distance h from the CM.

Solution:

(1) The axis locates at the CM. Take a small element of mass dm, as shown in Fig. 3-5,

$$dm = \lambda dx = \frac{m}{l}dx, \quad dI = x^2 dm = \lambda x^2 dx$$

$$I = \int dI = \int_{-l/2}^{l/2} \lambda x^2 dx = \frac{1}{3}\lambda x^3 \Big|_{-l/2}^{l/2} = \frac{1}{12}ml^2$$

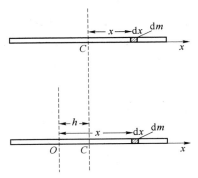

Fig. 3-5   Example 2

(2) The axis locates at arbitrary distance h from the CM.

$$I = \int_{-(l/2-h)}^{l/2+h} \lambda x^2 dx = \frac{1}{3}\lambda x^3 \Big|_{-l/2+h}^{l/2+h} = \frac{1}{12}ml^2 + mh^2$$

——The Parallel-axis theorem

3. Two blocks of masses m_A and m_B are connected by a light cord running over a pulley. The pulley is considered as a uniform cylindrical disk of mass m_C and radius R. There is no sliding between the pulley and the cord. There is frictionless between block A and horizontal surface. Find the acceleration of two blocks and the tensions acting on A and B.

Solution: Draw free-body diagrams. As shown in Fig. 3-6. The positive direction of rotation is clockwise. List Newton's Second Law for every object.

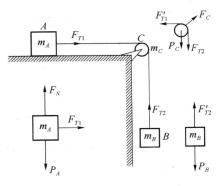

Fig. 3-6   Example 3

$$F_{T1} = m_A a$$

$$(F_{T2} - F_{T1})R = \left(\frac{1}{2}m_C R^2\right)\alpha$$

$$m_Bg - F_{T2} = m_Ba$$

There are 4 unknowns. The restriction condition: no sliding between the pulley and the cord, so there is $a = R\alpha$. We can obtain

$$a = \frac{m_Bg}{m_A + m_B + \frac{1}{2}m_C}, \quad F_{T1} = \frac{m_Am_Bg}{m_A + m_B + \frac{1}{2}m_C}, \quad F_{T2} = \frac{m_B\left(m_A + \frac{m_C}{2}\right)g}{m_A + m_B + \frac{1}{2}m_C}$$

4. A uniform rod of mass m and length l can pivot freely (no friction on the pivot) about a hinge to the ceiling. The rod is held horizontally and released. At time t, the angle between the rod and horizontal axis is θ, as shown in Fig. 3-7. Determine: (1) The angular acceleration and angular velocity of the rod as the function of θ; (2) The force on the hinge exerted by the rod.

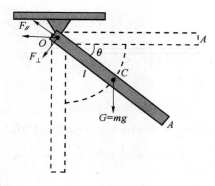

Fig. 3-7   Example 4

Solution:

(1) **Solution 1: Using Newton's second Law for the rotation of rod**

$$\frac{l}{2}mg\cos\theta = I\alpha = \left(\frac{1}{3}ml^2\right)\alpha$$

So the angular acceleration of the rod is $\alpha = \frac{3}{2}\frac{g}{l}\cos\theta$

According to the definition of the angular acceleration, we get

$$\alpha = \frac{d\omega}{dt} = \frac{d\omega}{d\theta}\frac{d\theta}{dt} = \omega\frac{d\omega}{d\theta} = \frac{3}{2}\frac{g}{l}\cos\theta$$

Then by using integral, we get $\int_0^\omega \omega d\omega = \frac{3}{2}\frac{g}{l}\int_0^\theta \cos\theta d\theta$

The angular velocity of the rod is obtained, $\omega = \sqrt{\frac{3g}{l}\sin\theta}$.

Solution 2: Using the law of conservation of mechanical energy

Take the Earth and rod as a system, the external force does not work and internal forces are constructive, so the mechanical energy of the system is conservative. So there is

$$0 = \frac{1}{2}\left(\frac{1}{3}ml^2\right)\omega^2 + \left(-mg\frac{l}{2}\sin\theta\right)$$

Solve the equation, we get the angular velocity of the rod at any position, $\omega = \sqrt{\frac{3g}{l}\sin\theta}$.

Chapter 3 Rotational Motion and Rigid Body

Using the definition of angular acceleration, we get

$$\alpha = \frac{d\omega}{dt} = \frac{d}{d\theta}\left(\sqrt{\frac{3g}{l}\sin\theta}\right)\frac{d\theta}{dt} = \sqrt{\frac{3g}{l}}\frac{\cos\theta}{2\sqrt{\sin\theta}}\sqrt{\frac{3g}{l}\sin\theta} = \frac{3g}{2l}\cos\theta$$

(2) Using Newton's Second Law for the CM of the rod

Normal: $F_{/\!/} - mg\sin\theta = ma_{n\text{-}CM} = m\dfrac{l}{2}\omega^2$

Tangential: $F_\perp + mg\cos\theta = ma_{t\text{-}CM} = m\dfrac{dv_{CM}}{dt} = m\dfrac{l}{2}\dfrac{d\omega}{dt} = m\dfrac{l}{2}\alpha$

Combine the (1) results, we get $F_\perp = -\dfrac{1}{4}mg\cos\theta$, $F_{/\!/} = \dfrac{5}{2}mg\sin\theta$

The force on the hinge exerted by the rod is $\boldsymbol{F} = -\dfrac{1}{4}mg\cos\theta\,\hat{\boldsymbol{\tau}} + \dfrac{5}{2}mg\sin\theta\,\hat{\boldsymbol{n}}$

5. The banging of a door against its stop can tear loose the hinges. By the proper choice of l, the impact forces on the hinge can be made to vanish. Determine the l. Suppose the moment of inertia of door is I and mass m.

Solution: Draw free-body diagrams, as shown in Fig. 3-8. The forces on the door during impact are F_d, due to the stop, and F' and F'' due to the hinge. F'' is the small radial force which provides the centripetal acceleration of swinging door. F' and F_d are the large impact forces which bring the door to rest when it bangs against the stop. To minimize the stress on the hinges, we must make F' as small as possible.

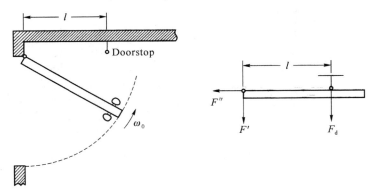

Fig. 3-8 Example 5

During the impact: Using Torque-Angular Momentum Theorem

$$L_f - L_i = \int_{t_i}^{t_f} M\,dt$$

$$L_i = I\omega_0, \quad L_f = 0, \quad M = -lF_d$$

So we get
$$I\omega_0 = l\int_{t_i}^{t_f} F_d\,dt \tag{3-14}$$

For the CM, using Impulse-Momentum Theorem in y-direction

$$p_{y\text{-}f} - p_{y\text{-}i} = \int_{t_i}^{t_f} F_y\,dt$$

$$p_{y\text{-}i} = mv_y = ml'\omega_0, \quad p_{y\text{-}f} = 0, \quad F_y = -(F' + F_d)$$

So we get
$$ml'\omega_0 = \int_{t_i}^{t_f} (F' + F_d) dt \tag{3-15}$$

Combine (3-14) and (3-15): $\int_{t_i}^{t_f} F' dt = \left(ml' - \dfrac{I}{l}\right)\omega_0$

When $F'=0$, the impact forces on the hinge can be made to vanish, according to $l = \dfrac{I}{ml'}$.

If the door is uniform and of width a, $I = \dfrac{1}{3}ma^2$ and $l' = \dfrac{a}{2}$, $\Rightarrow l = \dfrac{2}{3}a$.

The distance l specified by: $l = \dfrac{I}{ml'}$ is called "the center of percussion" or "sweet spot".
In batting a baseball, it is important to hit the ball at the bat's center of percussion to avoid a reaction on batter's hands and a painful sting, as shown in Fig. 3-9.

6. A circular platform of mass m and radius R rotates initially at an angular velocity ω_0 about its central axis. Then the platform is placed on a rough horizontal surface, as shown in Fig. 3-10. Determine (1) the torque acting on the platform by the friction force; (2) the time before the platform comes to a halt. The coefficient of friction between the platform and the surface is μ.

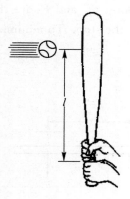

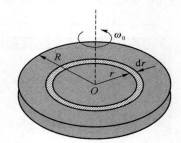

Fig. 3-9 Batting a baseball 　　　　Fig. 3-10 Example 6

Solution:

(1) The friction force is distributed in the whole area of the platform. Divide the whole platform into many circular rings with a radius of r and width dr, $dm = \sigma dS = \sigma 2\pi r dr$ and $dF_f = \mu g dm$

$$dM_f = -r dF_f = -\mu r g dm$$

The whole torque acting on the platform by the friction force is

$$M_f = -\int_m \mu r g\, dm = -\int_0^R \mu g r \sigma 2\pi r dr = -\dfrac{2}{3}\pi\mu g R^3 \sigma = -\dfrac{2}{3}\mu m g R$$

(2) Using the Newton's Second Law for rotation: $M_f = I\alpha$

$$-\dfrac{2}{3}\mu m g R = \dfrac{1}{2}mR^2 \dfrac{d\omega}{dt}$$

$$t = \int_0^t dt = -\dfrac{3R}{4\mu g}\int_{\omega_0}^0 d\omega = \dfrac{3R}{4\mu g}\omega_0$$

Chapter 3 Rotational Motion and Rigid Body

7. A rod of mass m' and length l can rotate about pivot O freely, a bullet of mass m and speed v_0 is shot into the lower end of the rod and embedded in the rod, as shown in Fig. 3-11. What is the angle θ when the rod swings to its highest position?

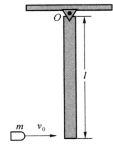

Fig. 3-11 Example 7

Solution: Take the bullet and the rod as a system. The external forces: the constraint force exerted by the pivot and gravity. They go through the origin O. So the external torque about O is zero, and the angular momentum of the system should be conserved in the process of shooting.

$$lmv_0 = \left(\frac{1}{3}m'l^2 + ml^2\right)\omega$$

$$\omega = \frac{3mv_0}{(m'+3m)l}$$

Take the bullet, the rod and the Earth as a system. In the process of the system swinging up, the mechanical energy is conserved.

$$\frac{1}{2}\left(\frac{1}{3}m'l^2 + ml^2\right)\omega^2 = mgl(1-\cos\theta) + m'g\frac{l}{2}(1-\cos\theta)$$

So we get $\cos\theta = 1 - \dfrac{3m^2}{(m'+3m)(m'+2m)}\dfrac{v_0^2}{gl}$

Questions and Problems

1. Two different disks of radius $r_1 > r_2$ are free to spin separately about an axis through the center and perpendicular to the plane of each disk. Both disks start from rest, and both undergo the same angular acceleration for the same length of time. Which disk will have the larger final angular velocity? _____.

　　(A) Disk 1

　　(B) Disk 2

　　(C) The disks will have the same angular velocity

　　(D) The answer depends on the mass of the disks

2. The pulley A and B are same. The light rope is wrapped around the pulley and does not slip over it. The pulley can rotate without friction about its axis. We tie the free end of the rope of pulley A to an object of mass M and at the same time act force F on the free end of the rope of pulley B, here $F = Mg$, as shown in Fig. 3-12. The angular acceleration of pulley A and B are β_A and β_B respectively. Then β_A and β_B satisfy _____.

　　(A) $\beta_A = \beta_B$ 　　　　　　　　(B) $\beta_A > \beta_B$

　　(C) $\beta_A < \beta_B$ 　　　　　　　　(D) At beginning $\beta_A = \beta_B$, then $\beta_A < \beta_B$

3. A horizontal circular platform can rotate without friction about the fixed perpendicular axis through its center. A child stands on it. At the beginning, the system of platform and child are at rest. Then the child starts to walk randomly on it. During the whole procedure for the system, which quantities are conserved? _____.

(A) Only the momentum is conserved
(B) Only the mechanical energy is conserved
(C) Only the angular momentum about the rotational axis is conserved
(D) Momentum, mechanical energy and angular momentum about the rotational axis are all conserved

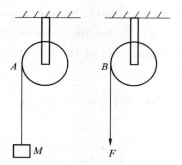

Fig. 3-12　Problem 2

4. A uniform rod of length $2L$ and mass m rests on a frictionless horizontal surface. The rod can rotate without friction about a fixed axis through its center and perpendicular to the surface. Two bullets (each bullet with mass m) traveling parallel to the horizontal surface and perpendicular to the rod with the same speed v but opposite direction struck the rod at two ends respectively and embed it, as shown in Fig. 3-13. The angular speed of the system after collision is _____.

(A) $\dfrac{2v}{3L}$ (B) $\dfrac{4v}{5L}$ (C) $\dfrac{6v}{7L}$

(D) $\dfrac{8v}{9L}$ (E) $\dfrac{12v}{7L}$

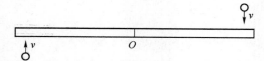

Fig. 3-13　Problem 4

5. A body, not necessarily rigid, is originally rotating with angular velocity of magnitude ω_0 and angular momentum of magnitude L_0. Sometimes happens to the body to cause ω_0 to slowly decrease. Consequently _____.

(A) L_0 must also be decreasing
(B) L_0 could be constant or decreasing, but not increasing
(C) L_0 could be constant, decreasing, or increasing
(D) L_0 could be constant or increasing, but not decreasing

6. A solid object is rotating without experiencing any external torques. In this case

(A) both the angular momentum and angular velocity have constant directions
(B) the direction of the angular momentum is constant but the direction of the

angular velocity might not be constant

(C) the direction of the angular velocity is constant but the direction of the angular momentum might not be constant

(D) neither the angular momentum nor angular velocity has a constant direction

7. A physical professor is sitting on a rotating chair with her arms outstretched, each holding a medium sized barbell. The frictionless chair is originally rotating at a constant speed. She then pulls her arms closer to her body.

(1) When she brings her arms in, her angular velocity _____.

(A) increases

(B) remains constant

(C) decreases

(D) changes, but whether it decreases or increases depends on how she brings her arms in

(2) When she brings her arms in, her angular momentum _____.

(A) increases

(B) remains constant

(C) decreases

(D) changes, but whether it decreases or increases depends on how she brings her arms in

8. Four solid objects, each with the same mass and radius, are spinning freely with the same angular speed. Which object requires the most work to stop it? _____.

(A) A solid sphere spinning about a diameter

(B) A hollow sphere spinning about a diameter

(C) A solid disk spinning about an axis perpendicular to the plane of the disk and through the center

(D) A hoop spinning about an axis along a diameter

(E) The work required is the same for all four objects

9. The flywheel in a car engine is under test. The angular position θ of the flywheel is given by $\theta = (2.0 \text{ rad/s}^3) t^3$, the diameter of the flywheel is 0.36 m. (1) The angle θ_1 at $t_1 = 2.0$ s is _____ and the angle θ_2 at $t_2 = 5.0$ s is _____; (2) The distance that a particle on the rim moves during that time interval is _____; (3) The average angular velocity between $t_1 = 2.0$ s and $t_2 = 5.0$ s is _____; (4) The instantaneous angular velocity at time $t = 3.0$ s is _____.

10. A bicycle wheel is being tested at a repair shop. The angular velocity of the wheel is 4.00 rad/s at time $t = 0$, and its angular acceleration is constant and equal to -1.20 rad/s^2. A spoke OP on the wheel coincides with the $+x$-axis at time $t = 0$. (1) The wheel's angular velocity at $t = 3.00$ s is _____; (2) At $t = 3.00$ s, the angle between the spoke OP and the $+x$-axis is _____.

11. A light, flexible, non-stretching cable is wrapped several times around a winch drum, a solid cylinder of mass 50 kg and diameter 0.120 m, which rotates about a stationary

horizontal axis held by frictionless bearings. The free end of the cable is pulled with a constant force of magnitude 9.0 N for a distance of 2.0 m, as shown in Fig. 3-14. Assuming that the cable unwinds without stretching or slipping, the angular acceleration of the cylinder is _____. If the cylinder is initially at rest, its final angular speed is _____ and the final speed of the cable is _____.

12. In a lab experiment to test conservation of energy in rotational motion, we wrap a light, flexible cable around a solid cylinder with mass M and radius R. The cylinder rotates with negligible friction about a stationary horizontal axis (Fig. 3-15). We tie the free end of the cable to an object of mass m and release the object with no initial velocity at a distance h above the floor. As the object falls, the cable unwinds without stretching or slipping, turning the cylinder. The speed of the falling object of the cylinder is _____ just as the object strikes the floor and the angular speed of the cylinder is _____ at that time.

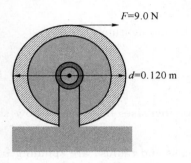
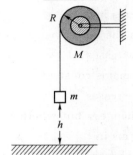

Fig. 3-14 Problem 11   Fig. 3-15 Problem 12

13. Fig. 3-16 shows a hollow, uniform cylinder with mass M, length L, inner radius R_1, and outer radius R_2. This might be a steel cylinder in a printing press or a sheet-steel rolling mill. The moment of inertia about the axis of symmetry of the cylinder is _____.

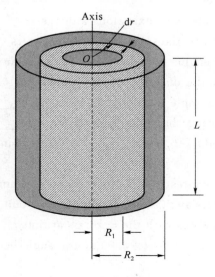

Fig. 3-16 Problem 13

Chapter 3 Rotational Motion and Rigid Body

14. The pulley in Fig. 3-17 has radius R and moment of inertia I. The rope does not slip over the pulley, and the pulley spins on a frictionless axle. The coefficient of kinetic friction between block A and the tabletop is μ_k. The system is released from rest, and block B descends. Block A has mass m_A and block B has mass m_B. The speed of block B as a function of the distance d that it has descended is _____ .

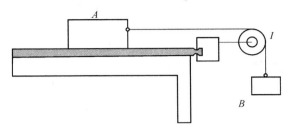

Fig. 3-17 Problem 14

15. A primitive yo-yo is made by wrapping a string several times around a solid cylinder with mass M and radius R. You hold the end of the string stationary while releasing the cylinder with no initial motion. The string unwinds but does not slip or stretch as the cylinder drops and rotates. The downward acceleration of the cylinder is _____ and the tension in the string is _____ . After it has dropped a distance h, as shown in Fig. 3-18, the speed v_{CM} of the center of mass of the solid cylinder is _____ .

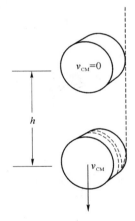

Fig. 3-18 Problem 15

16. Fig. 3-19 shows two disks, one is an engine flywheel, the other is a clutch plate attached to a transmission shaft. Their moments of inertia are I_A and I_B; initially, they are rotating with constant angular velocities ω_A and ω_B respectively. We then push the disks together with forces acting along the axis, so as not to apply any torque on either disk. The disks rub against each other and eventually reach a common final angular speed ω. The expression for ω is _____ .

17. An electric motor exerts a constant torque of $\tau = 100$ N·m on a grindstone mounted on its shaft. The moment of inertia of the grindstone is $I = 2.0$ kg·m². If the system starts from rest, the work done by the motor in 8.0 s is _____ and the kinetic energy at the end of this time is _____ . The average power delivered by the motor is _____ .

18. A door with 1.0 m wide, of mass 15 kg, is hinged at one side so that it can rotate without friction about a vertical axis. It is unlatched. A police officer fires a bullet with a mass of 10 g and a speed of 400 m/s into the exact center of the door, in a direction perpendicular to the plane of the door (Fig. 3-20). Just after the bullet imbeds itself in the door, the angular speed of the door is _____ , the kinetic energy of the system is _____ .

· 49 ·

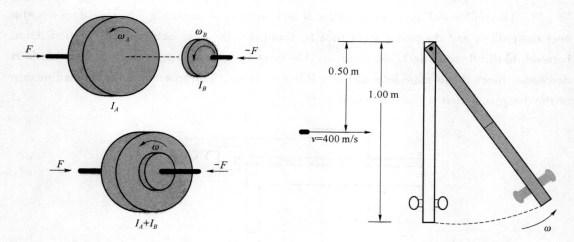

Fig. 3-19 Problem 16   Fig. 3-20 Problem 18

19. A small block on a frictionless horizontal surface has a mass of 0.025 kg. It is attached to a light cord passing through a hole in the surface (as shown in Fig. 3-21). The block is originally revolving at a distance of 0.300 m from the hole with an angular speed of 1.75 rad/s. The cord is then pulled from below, shortening the radius of the circle in which the block revolves to 0.150 m. The block may be treated as a particle. The new angular speed of the block is _____. The change in kinetic energy of the block is _____. The work was done in pulling the cord is _____.

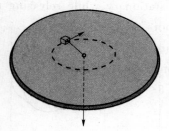

Fig. 3-21 Problem 19

20. A rigid object rotating about the z axis is slowing down at 2.66 rad/s². Consider a particle located at $r = (1.83 \text{ m})\, \hat{j} + (1.26 \text{ m})\, \hat{k}$. At the instant that $\omega = (14.3 \text{ rad/s})\, \hat{k}$, find (1) the velocity of the particle and (2) its acceleration. (3) What is the radius of the circular path of the particle?

21. A uniform beam of length L whose mass m is 1.8 kg rests with its ends on two digital scales, as shown in Fig. 3-22. A block whose mass M is 2.7 kg rests on the beam, its center one-fourth of the way from the beam's left end. What do the scales read?

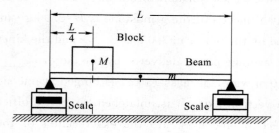

Fig. 3-22 Problem 21

22. A solid cylinder of mass M and radius R starts from rest and rolls without slipping

Chapter 3 Rotational Motion and Rigid Body

down an inclined plane of length L and height h, as shown in Fig. 3-23. Find the speed of its center of mass when the cylinder reaches the bottom.

23. A toy yo-yo of total mass $M=0.24$ kg consists of two disks of radius $R=2.8$ cm connected by a thin shaft of radius $R_0=0.25$ cm, as in Fig. 3-24. A string of length $L=1.2$ m is wrapped around the shaft. If the yo-yo is thrown downward with an initial velocity $v_0=1.4$ m/s, what is its rotational velocity when it reaches the end of the string?

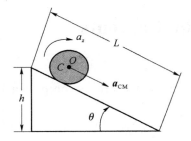

Fig. 3-23 Problem 22

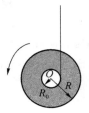

Fig. 3-24 Problem 23

24. A circular platform of mass m_0 and radius R rotates friction-free about an axis through its center. A woman with mass m stands on the platform at a distance $R/2$ from the center. At the beginning, the system of platform and woman rotates at the angular velocity ω_0 about the axis. The woman starts to walk to the edge of the platform, as shown in Fig. 3-25. Determine the final angular velocity ω of the system when the woman arrives at the edge.

25. A heavy steel chain of mass m and length l passes over a pulley of mass m_0 and radius r. The pulley is fixed with a frictionless pivot O. There is no slide between the chain and pulley. At the beginning, the chain passes over the pulley with the lengths of both sides equal. And then with a small perturbation, the chain slides to the left. Find the velocity and acceleration of the chain when the height difference of two ends is s, as shown in Fig. 3-26.

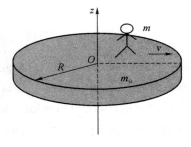

Fig. 3-25 Problem 24

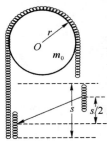

Fig. 3-26 Problem 25

Chapter 4   Electrostatics

Review of the Contents

1. Coulomb's law and electric field

In Fig. 4-1, the electrostatic force exerted by point charge q_1 on q_2 can be expressed in vector form as:

$$\boldsymbol{F}_{21} = k_e \frac{q_1 q_2}{r^2} \hat{\boldsymbol{r}}_{12} = \frac{1}{4\pi\varepsilon_0} \frac{q_1 q_2}{r^2} \hat{\boldsymbol{r}}_{12} \qquad (4\text{-}1)$$

where k_e is a proportionality constant, $\varepsilon_0 = 8.854,2 \times 10^{-12}$ C^2/(N·m^2) is the permittivity of free space, and $\hat{\boldsymbol{r}}_{12}$ is a unit vector directed from q_1 toward q_2.

From Newton's third law of motion, we can conclude that $\boldsymbol{F}_{21} = -\boldsymbol{F}_{12}$. The Coulomb forces between the two charges with opposite sign are attractive, while the two charges having the same sign result in repulsive Coulomb forces.

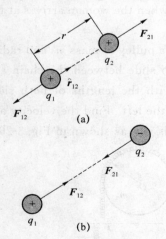

Fig. 4-1 The interaction between two point charges

It should be noted that Coulomb's law applies to objects whose size is much smaller than the distance between them. Actually, it is precise for point charges (spatial size is negligible). The law also describes the forces between two charges when they are at rest.

The definition of the electric field $\boldsymbol{E}$ in terms of the electric force exerted on a positive test charge q_0 placed at a particular point,

$$\boldsymbol{E} = \frac{\boldsymbol{F}_e}{q_0} \qquad (4\text{-}2)$$

$\boldsymbol{E}$ is the field produced by some charges or charge distribution separate from the test charge. It is not the field produced by the test charge itself. Therefore the presence of the test charge is not necessary for the field to exist. It serves as a detector of the electric field.

If the electric field is produced by a series of point charges in space, the total electric field at any point P obeys the superposition principle and equals the vector sum of the electric fields of all the charges,

$$\boldsymbol{E} = \sum_i \boldsymbol{E}_i = \frac{1}{4\pi\varepsilon_0} \sum_i \frac{q_i}{r_i^2} \hat{\boldsymbol{r}}_i \qquad (4\text{-}3)$$

If an object has a continuous charge distribution, using the superposition principle, the total electric field at any point P is,

$$\boldsymbol{E} = \int d\boldsymbol{E} = \frac{1}{4\pi\varepsilon_0} \int \frac{dq}{r^2} \hat{\boldsymbol{r}} \tag{4-4}$$

2. Gauss's law

Gauss's law is an alternative procedure for calculating electric fields. This law is based on the fact that the fundamental electrostatic force between point charges shows an inverse-square behavior. In practice, it is more convenient for calculating the electric fields of an object with highly symmetric charge distribution.

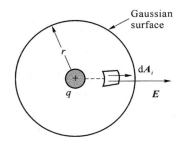

Fig. 4-2 A schematic of Gauss's law

The content of Gauss's law is that the net flux through any closed surface surrounding a point charge q is given by $\frac{q}{\varepsilon_0}$ and is independent of the shape of that surface, as shown in Fig. 4-2.

$$\oint_A \boldsymbol{E} \cdot d\boldsymbol{A} = \frac{q}{\varepsilon_0} \tag{4-5}$$

There is a point charge q and a spherical Gaussian surface with a radius of r.

To exploit Gauss's law, there are three key points to keep in mind.

(1) The flux only depends on the charges inside (enclosed).

(2) $\boldsymbol{E}$ on the left side of Gauss's law is referred to the $\boldsymbol{E}$ in Gaussian surface, and it is not necessarily due to the charge inside the surface. It is generated by all the charges in space.

(3) Zero flux doesn't mean the zero field.

3. Conductors in electrostatic equilibrium

Conductor is a kind of material in which electrons move freely. A conductor in electrostatic equilibrium has the following properties:

(1) The electric field is zero everywhere inside the conductor. If the field were not zero, free charges in the conductor would accelerate under the action of the electric field—not the case in electrostatic equilibrium.

(2) If the isolated conductor carries a net charge, the net charge resides entirely on its surface.

(3) The electric field just outside the charged conductor is perpendicular to the conductor surface and has a magnitude σ/ε_0, where σ is the surface charge density at that point.

$$\boldsymbol{E} = \frac{\sigma}{\varepsilon_0} \hat{\boldsymbol{n}} \tag{4-6}$$

(4) In Fig. 4-3, on an irregularly shaped conductor, the surface charge density is highest at locations where the curvature radius of the surface is smallest.

A Gaussian surface in the shape of a small cylinder is used to calculate the electric field just

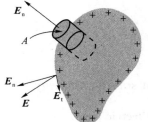

Fig. 4-3 A conductor with an arbitrary shape in electrostatic equilibrium

outside the conductor. The charge density is higher where the conductor has a larger curvature.

4. Electric potential

We define the electric potential difference ΔV to be the electric potential energy difference per unit test charge.

$$\Delta V = V_B - V_A = \frac{\Delta U}{q_0} = -\frac{1}{q_0}\int_A^B \boldsymbol{F} \cdot \mathrm{d}\boldsymbol{l} = -\int_A^B \boldsymbol{E} \cdot \mathrm{d}\boldsymbol{l} \tag{4-7}$$

The integration is performed along the path that q_0 follows as it moves from A to B, and the integral is called either a path integral or a line integral. Because the electrostatic force is a conservative force, this line integral does not depend on the path taken from A to B.

Electric potential at an arbitrary point in an electric field equals the work required per unit charge to bring a positive test charge from infinity to that point.

$$V_P = -\int_\infty^P \boldsymbol{E} \cdot \mathrm{d}\boldsymbol{l} \tag{4-8}$$

Electric potential is a scalar, independent of the charges that may be placed in the field. However, when we talk about electric potential energy, we are referring to the charge-field system.

5. Dielectric materials

A dielectric is a nonconducting material, such as rubber, glass, or waxed paper. Generally speaking, there are two types of dielectric materials, which are polar and nonpolar dielectric materials. When some dielectric material is inserted between the plates of a capacitor, the capacitance increases. This ability is represented by the dielectric constant. It is a property of a material and varies from one material to another.

When a dielectric material is placed in an external field $\boldsymbol{E}_0$, induced surface charges q' appear that tend to weaken the original field $\boldsymbol{E}_0$ by a polarization field $\boldsymbol{E}'$ within the material. For a linear material, the net field inside the material is $\boldsymbol{E} = \boldsymbol{E}_0 + \boldsymbol{E}'$. Induced charge q' is called bound charge. The bound charge resides in the surface of dielectric materials, that not free to move and bound to a molecule. The density of surface polarization charges is equal to the projection of polarization vector along normal direction of that surface.

$$\sigma' = \boldsymbol{P} \cdot \hat{\boldsymbol{n}} \tag{4-9}$$

For isotropic materials, $\boldsymbol{P} = \varepsilon_0 (\kappa - 1)\boldsymbol{E}$, where κ is the dielectric constant.

Typical Examples

Problem solving strategy

1. Electric field

In this chapter, the typical problem is to solve electric field distribution of charged objects. These theoretical methods are listed as follows:

(1) Coulomb's law and superposition principle.

Chapter 4 Electrostatics

The electric field due to a group of point charges can be solved by using the superposition principle. $\boldsymbol{E} = \sum_i \dfrac{q_i}{4\pi\varepsilon_0 r_i^2} \hat{\boldsymbol{r}}_i$.

The electric field due to a continuous charged body is $\boldsymbol{E} = \dfrac{1}{4\pi\varepsilon_0} \int \dfrac{dq}{r^2} \hat{\boldsymbol{r}}$, where dq is charge element, and r is the distance from charge element to the target point.

(2) Gauss's law. $\oint_S \boldsymbol{E} \cdot d\boldsymbol{A} = \dfrac{q_{in}}{\varepsilon_0}$. Gauss's law is very useful in obtaining electric fields when the charge distribution has a high degree of symmetry. The Gaussian surface should reflect the symmetry of the problem, in order to make the surface integration easier. For good choices of surface, the electric field is either parallel or perpendicular to the Gaussian surface.

(3) Potential gradient. $\boldsymbol{E} = -\nabla V$. This method means that if the electric potential distribution is known, the electric field can be obtained by derivation, and it represents the change of electric potential at a certain point in space.

2. Electric potential

(1) When dealing with a charge distribution of finite size, we usually define $V=0$ at an infinite far from the charged object.

(2) Divide the charge distribution into infinitesimal element of dq. Treat one element as a point charge, and define dV, $dV = \dfrac{dq}{4\pi\varepsilon_0 r}$.

(3) Obtain the total potential at P by integrating dV over the entire charge distribution $V = \int \dfrac{dq}{4\pi\varepsilon r}$.

(4) We can obtain the electric potential at P using another way in which Eq. (4-8) is used $V_P = -\int_\infty^P \boldsymbol{E} \cdot d\boldsymbol{l}$.

Examples

1. A rod of length l has a uniform positive charge per unit length λ and a total charge Q, as shown in Fig. 4-4. Calculate the electric field at P that is located along the long axis of the rod and a distance a from one end.

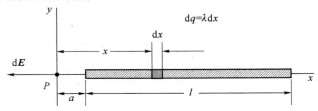

Fig. 4-4 A rod of length l with a uniform positive charge distribution

Solution: Suppose that dx is the length of one small segment, and the charge dq on the small segment is $dq = \lambda dx$.

The field due to this segment at P is the negative x direction, and its magnitude is

$$dE = \dfrac{dq}{4\pi\varepsilon_0 x^2} = \dfrac{\lambda dx}{4\pi\varepsilon_0 x^2}$$

The total field at P due to this rod is given by
$$E = \int_a^{l+a} \frac{dq}{4\pi\varepsilon_0 x^2} = \frac{Q}{4\pi\varepsilon_0 a(a+l)}$$

2. A thin rod bent into the shape of an arc of a circle of radius R carries a uniform charge (Fig. 4-5). Find the electric field along the axis of the rod starting at one end.

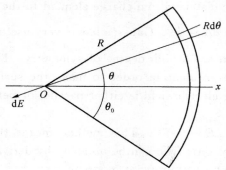

Fig. 4-5 A thin rod with an arc shape of radius of R

Solution: Select a differential element of the arc, the charge is $dQ = \lambda R d\theta$. The electric field produced by this element is $dE = \dfrac{dQ}{4\pi\varepsilon_0 R^2}$.

From the symmetry, the total field will have only an x-component.
$$E = \frac{\lambda R}{4\pi\varepsilon_0} \int_{-\theta_0}^{\theta_0} \frac{d\theta}{R^2}(-\cos\theta)$$
$$= \frac{-\lambda}{4\pi\varepsilon_0 R} \int_{-\theta_0}^{\theta_0} \cos\theta d\theta = \frac{-2\lambda\sin\theta_0}{4\pi\varepsilon_0 R}$$

3. An insulating solid sphere of radius a has a uniform volume charge density ρ and carries a total positive charge Q, as shown in Fig. 4-6. (1) Calculate the magnitude of the electric field at a point outside the sphere. (2) Find the magnitude of the electric field at a point inside the sphere.

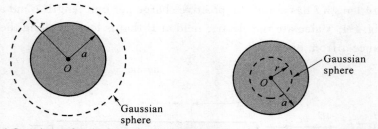

(a) A Gaussian surface is chosen to calculate the electric field of outside the sphere

(b) A Gaussian surface inside the sphere

Fig. 4-6 A solid spherical ball with a uniform charge density

Solution:

(1) Because the charge distribution is spherically symmetric, we select a spherical Gaussian surface of radius r. Applying Gauss's law,
$$\oint_A \boldsymbol{E} \cdot d\boldsymbol{A} = E \cdot 4\pi r^2 = \frac{Q}{\varepsilon_0}$$

$$E = \frac{Q}{4\pi\varepsilon_0 r^2}$$

(2) To determine the electric field inside the sphere, we select a spherical Gaussian surface. The volume of this smaller sphere is V'. To apply Gauss's law in this situation, the charge q_{in} within the Gaussian surface of volume V' is,

$$q_{in} = \rho V' = \rho\left(\frac{4}{3}\pi r^3\right)$$

From the property of symmetry, the magnitude of the electric field is constant everywhere on the spherical Gaussian surface and is normal to the surface at each point,

$$\oint_A \boldsymbol{E} \cdot d\boldsymbol{A} = E \cdot 4\pi r^2 = \frac{q_{in}}{\varepsilon_0}$$

$$E = \frac{\rho\left(\frac{4}{3}\pi r^3\right)}{4\pi r^2 \varepsilon_0} = \frac{Qr}{4\pi\varepsilon_0 a^3} \quad \text{(for } r<a\text{)}$$

4. An insulating solid sphere of radius R(Fig. 4-7) has a uniform positive volume charge density and total charge Q. (1) Find the electric potential at a point outside the sphere, that is, for $r>R$. Take the potential to be zero at $r=\infty$. (2) Find the potential at a point inside the sphere, that is, for $r<R$.

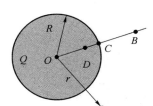

Fig. 4-7 A solid sphere of radius R with a uniform positive charge distribution

Solution:

The magnitude of the electric field inside a uniformly charged sphere of radius R is

$$\oint_A \boldsymbol{E} \cdot d\boldsymbol{A} = E \cdot 4\pi r^2 = \frac{1}{\varepsilon_0}\frac{r^3 Q}{R^3}, E = \frac{Qr}{4\pi\varepsilon_0 R^3}.$$

The outside electric field is $E = \frac{Q}{4\pi\varepsilon_0 r^2}$.

The electric potential at an exterior point B is

$$V_P = \int_B^\infty \boldsymbol{E} \cdot d\boldsymbol{r} = \int_r^\infty \frac{Q}{4\pi\varepsilon_0 r^2} dr = \frac{Q}{4\pi\varepsilon_0 r}.$$

The electric potential at an inside point D is

$$V_P = \int_D^\infty \boldsymbol{E} \cdot d\boldsymbol{r} = \int_r^R \frac{Qr}{4\pi\varepsilon_0 R^3} dr + \int_R^\infty \frac{Q}{4\pi\varepsilon_0 r^2} dr$$

$$= \frac{Q}{8\pi\varepsilon_0 R^3}(R^2 - r^2) + \frac{Q}{4\pi\varepsilon_0 R}$$

Questions and Problems

1. An imaginary, closed, spherical surface S of radius R is centered on the origin. A positive charge $+q$ is originally at the origin, and the flux through the surface is Φ_E. Three additional charges are now added along the x axis: $-3q$ at $x=-R/2$, $+5q$ at $x=R/2$, and $+4q$ at $x=3R/2$. The flux through S is now _____.

  (A) $2\Phi_E$ (B) $3\Phi_E$ (C) $6\Phi_E$ (D) $7\Phi_E$

2. If the potential is given by $V = xy - 3z^{-2}$, then the electric field has a y-component is _____.

(A) $-x$ (B) $-y$ (C) $-(x+y)$ (D) $x+y-6z^{-3}$

3. A parallel-plate capacitor is charged by connecting it to an ideal battery; the capacitor is then disconnected. Originally the energy stored in the capacitor is U_0. If the distance between the plates is doubled, then the new energy stored in the capacitor will be _____.

(A) $4U_0$ (B) $2U_0$ (C) U_0 (D) $U_0/2$

4. The electric field is uniform in Fig. 4-8. When a negative charge is moved from A to B, the electric potential energy of the charge _____ and the electric potential _____.

(A) increases; decreases   (B) remains constant; decreases

(C) decreases; increases   (D) increases; remains constant

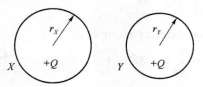

Fig. 4-8   A uniform electric field

5. Two conducting spheres, X and Y, have the same positive charge $+Q$, but different radii ($r_X > r_Y$) as shown in Fig. 4-9. The spheres are separated so that the distance between them is large compared with either radius. If a wire is connected between them, in which direction will current be directed in the wire? _____.

(A) From X to Y (B) From Y to X

(C) There will be no current in the wire

(D) It cannot be determined without knowing the magnitude of Q

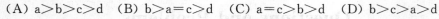

Fig. 4-9   Two spherical conductors with different radius and same charges

6. In Fig. 4-10, four circular plastic rods with uniform charge $+Q$ produce an electric field of magnitude E at the center of curvature (at the origin). Rank the four arrangements according to the magnitude of the electric field at the center of curvature, _____.

(A) $a > b > c > d$ (B) $b > a = c > d$ (C) $a = c > b > d$ (D) $b > c > a > d$

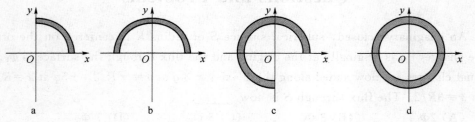

Fig. 4-10   Four charged plastic rod with different shapes

7. Charge is distributed uniformly throughout a long non-conducting cylinder of radius R. Which of the following graphs best represents the magnitude of the resulting electric field E as a function of r, the distance from the axis of the cylinder? _____.

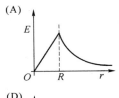
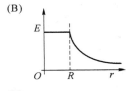
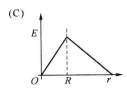
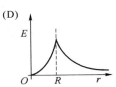
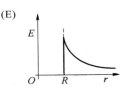

8. Positive charge Q is uniformly distributed over a thin ring of radius R that lies in a plane perpendicular to the x-axis, with its center at the origin O. Which of the following graphs best represents the electric field along the positive x-axis? _____.

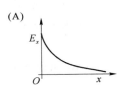
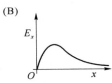
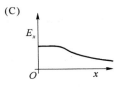
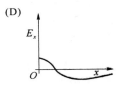

9. Consider Gauss's law: $\oint_A \mathbf{E} \cdot d\mathbf{A} = q/\varepsilon_0$. Which of the following is true? _____.

(A) $\mathbf{E}$ must be the electric field due to the enclosed charge

(B) If $q=0$ then $\mathbf{E}=0$ everywhere on the Gaussian surface

(C) If the charge inside consists of an electric dipole, then the integral is zero

(D) If a charge is placed outside the surface, then it cannot affect $\mathbf{E}$ on the surface

10. An infinite plane sheet is with a uniform surface charge density σ. The potential difference between the points a and b, both with distance h from the surface of the sheet is _____.

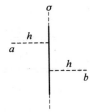

(A) 0

(B) $\dfrac{\sigma}{2\varepsilon_0}$

(C) $\dfrac{\sigma h}{\varepsilon_0}$

(D) $\dfrac{2\sigma h}{\varepsilon_0}$

11. Which of following is the right charge object which can produce the electric field in space with a spherical symmetry? The curve of E versus. r is shown in Fig. 4-11. _____.

(A) A uniformly charged thin sphere shell of radius R

(B) A uniformly charged solid sphere of radius R

(C) A solid sphere of radius R, with non-uniform charge density $\rho=Ar$, where A is a constant

(D) A solid sphere of radius R, with non-uniform charge density $\rho=A/r$

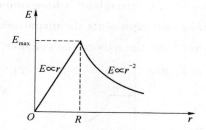

Fig. 4-11 The curve of E versus r of a charged object

12. A uniformly charged ring of radius R has a total charge Q, with $V=0$ at infinity, the electric potential at the center of the circle is _____.

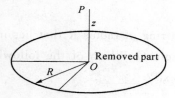

Fig. 4-12 A quadrant of charged plastic section

13. A plastic disk is charged on one side with a uniform surface charge density σ, and then three quadrants of the disk are removed. The remaining quadrant is shown in Fig. 4-12 with $V=0$ at infinity, the potential due to the remaining quadrant at point P, which is on the central axis of the original disk at a distance z from the original center, is _____.

14. A spherical capacitor, whose inner and outer radius is R_1 and R_2 separately, has potential difference V. The energy stored in this capacitor is _____.

15. A parallel-plate capacitor of plate area A is filled with two dielectrics as shown in Fig. 4-13. The capacitance is _____.

16. A point charge q is placed at the vertex of a cube as shown in Fig. 4-14. The electric flux through the surface $ABCD$ of the cube is _____.

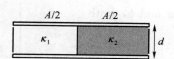

Fig. 4-13 A capacitor with two different dielectrics

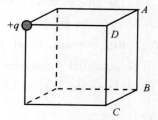

Fig. 4-14 A point charge at one vertex of a cube

17. Each plate of a parallel-plate capacitor exerts a force on the other can be expressed in terms of charge Q and area A as _____.

18. The electrostatic energy stored in the electric field outside an isolated spherical conductor of radius R carrying a net charge Q is _____.

19. Points A [at (2, 3) m] and B [at (5, 7) m] are in a region where the electric field is uniform and given by $E = 4\hat{i} + 3\hat{j}$ N/C. Then the potential difference $V_A - V_B$ should be _____.

Chapter 4　Electrostatics

20. The electric potential in a region is given by $V = 2xy + y^2 - 4xyz$, and then the electric field vector in this region should be _____.

21. The plastic rod shown in Fig. 4-15 has length L and a non-uniform linear charge density $\lambda = Ax$, where A is a positive constant. Find the electric field at point P on the axis, at distance b from one end.

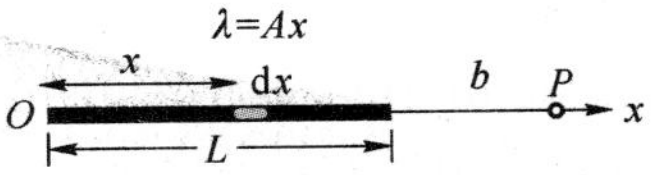

Fig. 4-15　A plastic rod with a varying charge density

22. In Fig. 4-16, a sphere, of radius a and charge $+q$ uniformly distributed throughout its volume, is concentric with a spherical conducting shell of inner radius b and outer radius c. This shell has a net charge of $-q$. Find expressions for the electric field, as a function of the radius r, (1) within the sphere ($r < a$), (2) between the sphere and the shell ($a < r < b$), (3) inside the shell ($b < r < c$), and (4) outside the shell ($r > c$). (5) What are the charges on the inner and outer surfaces of the shell?

23. In Fig. 4-17, a nonconducting spherical shell, of inner radius a and outer radius b, has a positive volume charge density $\rho = A/r$ (within its thickness), where A is a constant and r is the distance from center of the shell. Besides, a positive point charge q is located at the center. What value should A have if the electric field in the shell ($a \leqslant r \leqslant b$) is to be uniform? (Hint: The constant A depends on a but not on b)

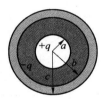
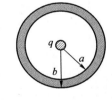

Fig. 4-16　A spherical object with a radius a and charge $+q$ and a conducting shell are concentric

Fig. 4-17　A point charge is located at the center of a spherical shell

24. A very long conducting cylindrical rod of length L with a total charge $+q$ is surrounded by a conducting cylindrical shell (also of length L) with total charge $-2q$, as shown in Fig. 4-18. Use Gauss's law to find (1) the electric field at points outside the conducting shell, (2) the distribution of charge on the shell, and (3) the electric field in the region between the shell and rod.

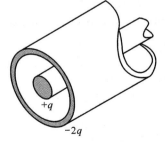

25. A parallel-plate capacitor has plates of area A and separation d and is charged to a potential difference V. The charging battery is then disconnected, and the plates are pulled apart until their separation is $2d$. Derive expressions in terms of A, d, and V for (1) the new potential difference; (2) the initial and final stored energies, U_i and U_f; and (3) the work required to separate the plates.

Fig. 4-18　A very long conducting cylindrical rod with a charge $+q$

26. A solid nonconducting sphere of radius R has a nonuniform charge distribution of

volume charge density $\rho = \dfrac{kr}{R}$, where k is a constant and r is the distance from the center of the sphere. (1) Show that the total charge of the sphere is $Q = \pi k R^3$. (2) Find the magnitude of the electric field both inside and outside the sphere. (3) Find the electric potential at the center of the sphere $V(0)$ if $V(\infty) = 0$.

27. A very large thin plane shown in Fig. 4-19 has uniform surface charge density σ. Touching it on the right is a long and wide slab of thickness d with uniform volume charge density ρ_E. Determine the electric field (1) to the left of plane, (2) to the right of the slab, and (3) everywhere inside the slab.

28. The flat disk of radius R (shown in Fig. 4-20) has a nonuniform surface charge distribution $\sigma = ar^2$. Determine the potential V at points along the x axis, relative to $V = 0$ at $x \to \infty$.

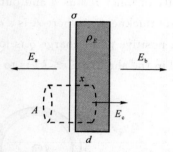
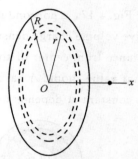

Fig. 4-19 A long and wide slab is touching with a charged thin plane

Fig. 4-20 A disk with a radius of R and charge density σ

29. Suppose one plate of a parallel-plate capacitor were tilted so it made a small angle θ with the other plate, as shown in Fig. 4-21. Determine a formula for capacitance C in terms of θ, where θ is small. Assume the plates are square. (Hint: Imagine the capacitor as many infinitesimal capacitors in parallel.)

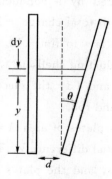

Fig. 4-21 A capacitor with a tilted plate

Chapter 5 Magnetism

Review of the Contents

1. The Biot-Savart law

The magnetic field described by the Biot-Savart law is the field due to a given current-carrying conductor, as shown in Fig. 5-1.

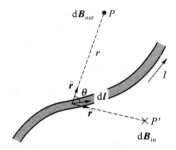

Fig. 5-1 The magnetic field d**B** due to the current I through a length element d**l** is obtained by the Biot-Savart law

The content of the Biot-Savart law is

$$d\boldsymbol{B} = \frac{\mu_0}{4\pi} \frac{I d\boldsymbol{l} \times \hat{\boldsymbol{r}}}{r^2} \tag{5-1}$$

In this law, d**B** is the field created by the current in only a small element d**l** of the conductor. The total magnetic field created at some point by a current of finite size is

$$\boldsymbol{B} = \int d\boldsymbol{B} = \frac{\mu_0}{4\pi} \int \frac{I d\boldsymbol{l} \times \hat{\boldsymbol{r}}}{r^2} \tag{5-2}$$

where the integral is taken over the entire current distribution.

The expression is developed from a current-carrying wire, and it is also valid for a current consisting of charges flowing through space.

2. Gauss's law in magnetism

In magnetism, we use magnetic field lines to describe magnetic field. Unlike electrostatic field, magnetic field lines are continuous and form closed loops. They do not begin or end at any points. The Gauss's law in magnetism is expressed as

$$\Phi = \oint_A \boldsymbol{B} \cdot d\boldsymbol{A} = 0 \tag{5-3}$$

In Fig. 5-2, the magnetic field lines of the bar magnet, for any closed surface, the number of lines entering the surface equals the number leaving the surface; thus the net magnetic flux is zero.

3. Ampère's law

In fact, it is impossible to solve the magnetic field of a general current-carrying conductor using the Biot-Savart law. Gauss's law in magnetism indicates that the magnetic field lines are always closed and it is always giving zero result of the surface integral of magnetic field. So it can't be used to determine the magnetic field produced by a particular current distribution.

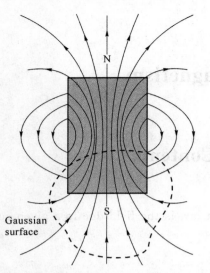

Fig. 5-2 A magnetic field of a bar magnet is shown by magnetic field lines

However, in order to calculate the magnetic field of current configurations having a high degree of symmetry, Ampère's law is used to describe the distribution of magnetic fields by these magnetic sources. The content of Ampère's law is that the line integral of $\boldsymbol{B} \cdot \mathrm{d}\boldsymbol{l}$ around any closed path equals $\mu_0 I$, where I is the total steady current passing through any surface bounded by the closed path.

$$\oint_l \boldsymbol{B} \cdot \mathrm{d}\boldsymbol{l} = \mu_0 I \tag{5-4}$$

The use of Ampère's law in magnetism is similar to that of Gauss's law in electrostatic field, which is capable of calculating electric fields for highly symmetric charge distributions.

4. Magnetic force

The essence of magnetic field is from moving charges. This field can exert a force on a current-carrying object or on a moving charge.

(1) Magnetic force acting on a current-carrying conductor

In Fig. 5-3, a current-carrying wire experiences a force when placed in a magnetic field. For an arbitrarily shaped wire segment of uniform cross-section in a magnetic field, the magnetic force exerted on a small segment of vector length $\mathrm{d}\boldsymbol{l}$ is

$$\mathrm{d}\boldsymbol{F} = I \mathrm{d}\boldsymbol{l} \times \boldsymbol{B} \tag{5-5}$$

where $\mathrm{d}\boldsymbol{F}$ is directed out of the page.

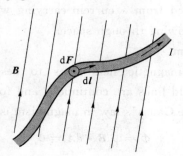

Fig. 5-3 A current segment of arbitrary shape carrying a current I in a magnetic field $\boldsymbol{B}$

The total force acting on the wire is the integral over the length of the wire:

$$F = I\int dl \times B \quad (5\text{-}6)$$

If we put a closed current-carrying loop in a uniform magnetic field, the net magnetic force on the loop is zero.

$$F = I\oint dl \times B = 0 \quad (5\text{-}7)$$

(2) Torque on a current loop in a uniform magnetic field

Consider a rectangular loop carrying a current I in a uniform magnetic field, as shown in Fig. 5-4, the net torque tends to rotate the loop and is defined by:

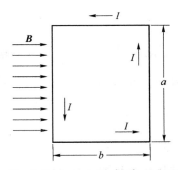

Fig. 5-4 A rectangular loop in a uniform magnetic field

$$M = (IA) \times B = \mu \times B \quad (5\text{-}8)$$

where μ is the magnetic dipole moment and A is equal to ab.

(3) Magnetic force on a moving charge

The magnetic force acting on a charged particle moving in a magnetic field is perpendicular to the velocity of the particle. This force is also called Lorentz force. The expression of magnetic force on a moving charging is

$$F = qv \times B \quad (5\text{-}9)$$

If the velocity of a charged particle is perpendicular to a uniform magnetic field, the particle moves in a circular path. However, if a charged particle moves in a uniform magnetic field with its velocity at some arbitrary angle with respect to B, its path is a helix.

5. The Hall effect

If an electric current flows through a conductor in a magnetic field, the magnetic field exerts a transverse force on the moving charge carriers which tends to push them to one side of the conductor. The presence of this measurable transverse voltage is called the Hall effect after E. H. Hall who discovered it in 1879.

This effect is most evident in a thin flat conductor as illustrated in Fig. 5-5. A magnetic field is applied to a current-carrying conductor. When I is in the x direction and B in the y direction, a buildup of charge at the sides of the conductors will balance this magnetic influence, producing a measurable voltage between the two sides of the conductor. The Hall voltage measured between points a and c is,

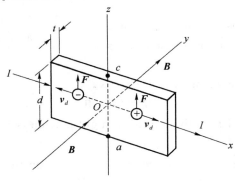

Fig. 5-5 A magnetic field is applied to a current-carrying conductor, in order to observe the Hall effect. In this configuration, both positive and negative charges are deflected upward in the magnetic field. The Hall voltage is measured between points a and c

$$\Delta V_H = \frac{IB}{nqt} = \frac{R_H IB}{t} \qquad (5\text{-}10)$$

where $R_H = \frac{1}{nq}$ is the Hall coefficient. It shows that a properly calibrated conductor can be used to measure the magnitude of an unknown magnetic field.

Typical Examples

Problem solving strategy

1. Magnetic field

In this chapter, the magnetic field of current-carrying objects can be obtained by the following ways.

(1) The Biot-Savart law. $\boldsymbol{B} = \int d\boldsymbol{B} = \frac{\mu_0}{4\pi} \int \frac{I d\boldsymbol{l} \times \hat{r}}{r^2}$. We can determine the magnetic field $d\boldsymbol{B}$ at a point due to the current I through a length element $d\boldsymbol{l}$, that is $d\boldsymbol{B} = \frac{\mu_0}{4\pi} \frac{I d\boldsymbol{l} \times \hat{r}}{r^2}$. The direction of the magnetic field is connected with the cross product of $d\boldsymbol{l} \times \hat{r}$, where $\hat{r}$ is a unit vector of position vector from that certain point to current element.

(2) Ampère's law. $\oint_l \boldsymbol{B} \cdot d\boldsymbol{l} = \mu_0 I$. Like Gauss's law in electrostatic field, Ampère's law is very effective in solving the magnetic field of symmetric steady current-carrying objects. To evaluate the line integral in Ampère's law, we should construct a closed path, which can reflect the symmetry of magnetic field.

Examples

1. In Fig. 5-6, a semicircle wire of radius R forms a closed circuit and carries a current I. The wire lies in the xy plane, and a uniform magnetic field is directed along the positive y axis. Find the magnetic force acting on the wire.

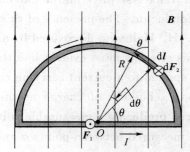

Fig. 5-6 Example 1. A closed circuit with a radius of R and a current I

Solution: Because a closed loop is put in a uniform magnetic field, the net magnetic force is zero.

For the straight portion, using Ampère's law,
$$d\boldsymbol{F} = I d\boldsymbol{l} \times \boldsymbol{B}$$
we get that the magnitude of F_1 acting on the straight portion is $F_1 = 2IRB$, and the direction is out of the page.

To find the force acting on the curved part, the force on the length element $d\boldsymbol{l}$ is,
$$dF_2 = I |d\boldsymbol{l} \times \boldsymbol{B}| = IB\sin\theta dl = IRB\sin\theta d\theta$$

The net force on the curved part is

$$dF_2 = IRB\int_0^\pi \sin\theta d\theta = 2IRB$$

The direction of F_2 is into the page.

2. In Fig. 5-7, a thin, infinitely large sheet lying in the yz plane carries a current of linear current density J. The current is in the y direction, and J represents the current per unit length measured along the z axis. Find the magnetic field near the sheet.

Solution: We can take a rectangular path through the sheet. Because the current distribution is uniform on the infinite sheet, by symmetry, we can conclude that the magnetic field is constant over the sides of length l, and hence the field should not vary from point to point.

Applying Ampère's law, we can get the magnetic field outside the infinite sheet,

$$\oint \boldsymbol{B} \cdot d\boldsymbol{l} = \mu_0 I = \mu_0 Jl$$

$$B = \mu_0 \frac{J}{2}$$

The direction of magnetic field is parallel to the surface of sheet.

3. A rectangular loop of width a and length b is located near a long wire as shown in Fig. 5-8. The distance between the wire and the closest side of the loop is c. The wire is parallel to the long side of the loop. Calculate the total magnetic flux through the loop due to the current in the wire.

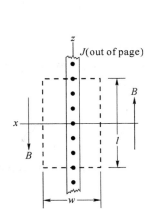

Fig. 5-7 Example 2

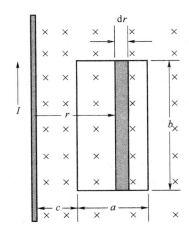

Fig. 5-8 Example 3

Solution: We know that the magnetic field created by an infinite wire is

$$B = \frac{\mu_0 I}{2\pi r}$$

To calculate the magnetic flux through the rectangular loop, we choose a small area element dA on the loop, the magnetic flux is

$$d\Phi_B = \boldsymbol{B} \cdot d\boldsymbol{A} = \frac{\mu_0 I}{2\pi r} dA$$

The total magnetic flux is the integral of the whole area,

$$\Phi_B = \int d\Phi_B = \frac{\mu_0 I}{2\pi}\int_c^{a+c} \frac{b\,dr}{r} = \frac{\mu_0 Ib}{2\pi}\ln\left(\frac{a+c}{c}\right)$$

Questions and Problems

1. What is the value of $\oint \boldsymbol{B} \cdot d\boldsymbol{l}$ for the path shown in Fig. 5-9? _____.

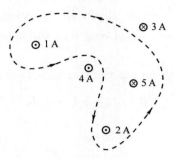

Fig. 5-9   Problem 1

(A) $-8\pi \times 10^{-7}$ T·m (B) $-4\pi \times 10^{-7}$ T·m
(C) $+8\pi \times 10^{-7}$ T·m (D) $+32\pi \times 10^{-7}$ T·m

2. Magnetization is _____.
   (A) the current density in an object
   (B) the charge density of moving charges in an object
   (C) the magnetic dipole moment of an object
   (D) the magnetic dipole moment per unit volume of an object

3. The negatively charged disk in Fig. 5-10 is rotating clockwise. What is the direction of the magnetic field at point A in the same plane of the disk? _____.
   (A) Into the page   (B) Out of the page
   (C) Up the page   (D) Down the page

4. Fig. 5-11 shows several wire segments that carry equal current from a to b. The wires are in a uniform magnetic field $\boldsymbol{B}$ directed into the page. Which wire segment experiences the largest net force? _____.
   (A) 1   (B) 2
   (C) 3   (D) All experience the same net force

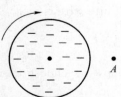

Fig. 5-10   Problem 3

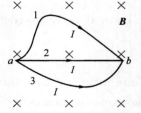

Fig. 5-11   Problem 4. The model of three wire segments in magnetic field

Chapter 5 Magnetism

5. A loop of wire of length L carrying a current I (Fig. 5-12) can be wound once as in the left, or twice in the right. The ratio of the magnetic field B_1 at the center of left single loop to the field B_2 at the center of the right double loop is _____ .

(A) 2 (B) 1
(C) 1/2 (D) 1/4

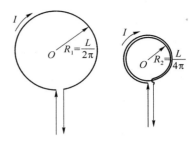

Fig. 5-12   Problem 5

6. A square conductor (Fig. 5-13) moves through a uniform magnetic field. Which of the figures shows the correct charge distribution on the conductor? _____ .

(A) (a) (B) (b) (C) (c)
(D) (d) (E) (e)

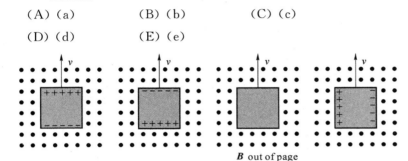

Fig. 5-13   Problem 6

7. Consider the hemispherical closed surface (Fig. 5-14). The hemisphere is in a uniform magnetic field that makes an angle θ with the vertical. What is the magnetic flux through the hemispherical surface S_2? _____ .

(A) $\pi r^2 B$ (B) $2\pi r^2 B$ (C) $\pi r^2 B \sin\theta$ (D) $-\pi r^2 B \cos\theta$

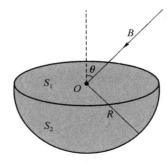

Fig. 5-14   Problem 7

8. The magnetic field at point A for a current carrying square wire in Fig. 5-15 is _____ .

(A) $\dfrac{\sqrt{2}\mu_0 I}{4\pi l}$ (B) $\dfrac{\sqrt{2}\mu_0 I}{2\pi l}$ (C) $\dfrac{\sqrt{2}\mu_0 I}{\pi l}$ (D) 0

9. Consider a circular current loop in a uniform magnetic field **B** in Fig. 5-16. Now suppose that a, b and c are small segment with equal length, which of the following is true? _____.

(A) $F_b > F_c > F_a$ (B) $F_a > F_b > F_c$ (C) $F_a < F_b = F_c$ (D) $F_a > F_c = F_b$

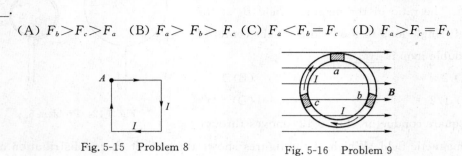

Fig. 5-15 Problem 8   Fig. 5-16 Problem 9

10. A very long, thin strip of copper of width w carries a current I along its length. The point P is in the plane of the strip at distance b away from it. What is the magnetic field at point P in Fig. 5-17? _____.

(A) $\dfrac{\mu_0 I}{2\pi(w+b)}$ (B) $\dfrac{\mu_0 I}{2\pi w}\ln\dfrac{w+b}{b}$ (C) $\dfrac{\mu_0 I}{2\pi b}\ln\dfrac{w+b}{b}$ (D) $\dfrac{\mu_0 I}{\pi(a+2b)}$

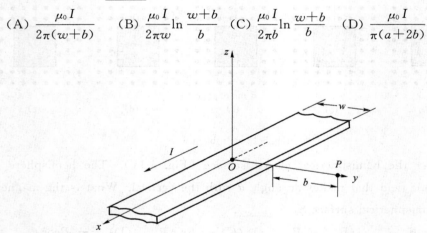

Fig. 5-17 Problem 10

11. Four long, parallel conductors carry equal currents of I. Fig. 5-18 is an end view of the conductors. The current direction is into page at points A and C (indicated by the crosses) and out of the page at B and D (indicated by the dots). What is the magnetic field at point O? _____.

(A) $B = \dfrac{2\mu_0}{\pi a}I$

(B) $B = \dfrac{\sqrt{2}\mu_0}{2\pi a}I$

(C) $B = \dfrac{\mu_0}{\pi a}I$

(D) $B = 0$

Fig. 5-18 Problem 11

12. Consider a long hollow cylindrical conductor having inner radius a and outer radius b. The current density is uniform over the cross section of conductor. Which of the diagram in Fig. 5-19 is correct for the magnetic field B at a distance $r>0$, measured from the axis. _____.

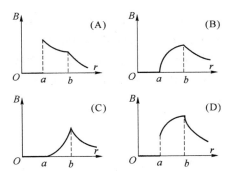

Fig. 5-19 Problem 12

13. A conductor consists of a circular loop of radius R and two straight, long sections (Fig. 5-20). The wire lies in the plane of the paper and carries a current I. What is the vector magnetic field at the center of the loop? _____.

(A) $\dfrac{\mu_0 I}{2\pi R}$ (B) $\dfrac{\mu_0 I}{2R}\left(1-\dfrac{1}{\pi}\right)$ (C) $\dfrac{\mu_0 I}{2R}\left(1+\dfrac{1}{\pi}\right)$ (D) $\dfrac{\mu_0 I}{4R}$

Fig. 5-20 Problem 13

14. Consider two identical closed circular paths around two long, straight current-carrying wires. In Fig. 5-21 (b), there is an extra long, straight current-carrying wire. Both are in a vacuum. Which of the following expressions is true? _____.

(A) $\oint_{L_1} \boldsymbol{B} \cdot d\boldsymbol{l} = \oint_{L_2} \boldsymbol{B} \cdot d\boldsymbol{l}$, $B_{P_1} = B_{P_2}$

(B) $\oint_{L_1} \boldsymbol{B} \cdot d\boldsymbol{l} \neq \oint_{L_2} \boldsymbol{B} \cdot d\boldsymbol{l}$, $B_{P_1} = B_{P_2}$

(C) $\oint_{L_1} \boldsymbol{B} \cdot d\boldsymbol{l} = \oint_{L_2} \boldsymbol{B} \cdot d\boldsymbol{l}$, $B_{P_1} \neq B_{P_2}$

(D) $\oint_{L_1} \boldsymbol{B} \cdot d\boldsymbol{l} \neq \oint_{L_2} \boldsymbol{B} \cdot d\boldsymbol{l}$, $B_{P_1} \neq B_{P_2}$

Fig. 5-21 Problem 14

15. An electron moves through a uniform magnetic field given by $\boldsymbol{B} = a\,\hat{\boldsymbol{i}} + 3a\hat{\boldsymbol{j}}$. At a particular instant, the electron has the velocity $\boldsymbol{v} = 2.0\,\hat{\boldsymbol{i}} + 4.0\,\hat{\boldsymbol{j}}$ and the magnetic force acting on it is $(6.4 \times 10^{-19})\hat{\boldsymbol{k}}$, all quantities are in SI units. The constant $a =$ _____.

16. Fig. 5-22 shows five possible orientations of a magnetic dipole $\boldsymbol{\mu}$ in a uniform magnetic field $\boldsymbol{B}$. The greatest magnitude of the magnetic torque on the dipole $\boldsymbol{\mu}$ among these should be _____.

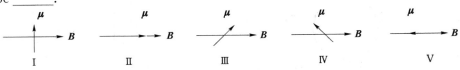

Fig. 5-22 Problem 16

17. Four long straight wires carry equal currents into the page as shown in Fig. 5-23. The magnetic force exerted on wire F is in the direction of _____.

18. As shown in Fig. 5-24, uniform electric current flows into and then out from a circular ring. The magnetic field at the center of the ring should be _____.

Fig. 5-23 Problem 17   Fig. 5-24 Problem 18

19. A long wire is bent into the shape as shown in Fig. 5-25. (All parts are in the same plane of the paper and part 2 is a quarter of a circular arc of radius R). Calculate magnetic field $\boldsymbol{B}$ at the point O.

20. A wire is formed into the shape of two half circles connected by equal-length straight sections as shown in Fig. 5-26(current I is clockwise). (1) What are the magnitude and direction of magnetic field $\boldsymbol{B}$ at the center C? (2) Find the magnetic dipole moment μ of the circuit(give both the magnitude and the direction of μ).

Fig. 5-25 Problem 19   Fig. 5-26 Problem 20

21. A coil of N turns is closely wound along the surface of a half ball (radius R). Each turn carries a current I. All coils cover the whole semi-circle parallel. Find the magnetic field at the center O of the ball.

22. The circular wire loop of radius R in Fig. 5-27 carries a clockwise current I. There is a uniform magnetic field B directed to the right. (1) What is the net force on the current loop $\boldsymbol{B}$? (2) Find the torque $\boldsymbol{M}$ on the current loop.

23. Suppose a nonconducting rod of length l carries a uniformly distributed charge Q as shown in Fig. 5-28. It is rotated with angular velocity ω about an axis perpendicular to the rod at one end. What is the magnetic dipole moment of this rod?

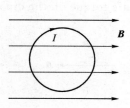   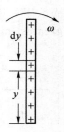

Fig. 5-27 Problem 22   Fig. 5-28 Problem 23

24. A very long flat conducting strip of wide L and negligible thickness lies in a horizontal plane and carries a uniform current I across its cross section, as shown in Fig. 5-29 (1) Show that at point A a distance y directly above its center, the field is given by $B = \dfrac{\mu_0 I}{\pi L} \arctan \dfrac{L}{2y}$, assuming the strip is infinitely long. (2) What value does B approach for $y \gg L$?

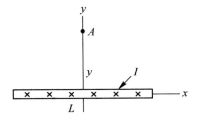

Fig. 5-29 Problem 24

25. A circular conducting ring of radius R is connected to two exterior straight wires ending at two ends of a diameter. The current I splits into unequal portions while passing through the ring as shown in Fig. 5-30, what is B at the center of the ring?

26. A wire is bent into the shape of a regular polygon with n sides whose vertices are a distance R from the center in Fig. 5-31, if the wire carriers a current I_0, (1) determine the magnetic field at the center; (2) if n is allowed to become very large ($n \to \infty$), show that the formula in part (1) reduces to that for a circular loop.

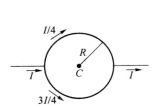 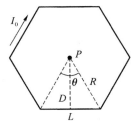

Fig. 5-30 Problem 25   Fig. 5-31 Problem 26

Chapter 6　Electromagnetic Inductance

Review of the Contents

1. Faraday's law of induction

The magnetic field can be created by moving charges or electric current. In 1831, Michael Faraday verified that an electromotive force (emf) can be induced in a circuit by a changing magnetic field. The results led to a very basic and very important law of electromagnetism known as Faraday's law of induction, as shown in Fig. 6-1. Its content is that the magnitude of the emf induced in a circuit equals the time rate of change of the magnetic flux through the circuit. The expression of Faraday's law of induction is:

$$\varepsilon = -\frac{d\Phi_B}{dt} \tag{6-1}$$

where Φ_B is the magnetic flux through the circuit. In this law, "$-$" means that the induced current is in a direction such that the induced magnetic field attempts to maintain the original flux through the loop. It is related to Lenz's law.

From Faraday's law, we can conclude that an emf can be induced in the circuit in following ways:
- The magnitude of B can change with time.
- The area enclosed by the loop can change with time.
- The angle between B and the normal to the loop can change with time.

Fig. 6-1　The illustration of Faraday's law of induction

2. Motional emf

Motional emf is induced when a conductor is moving through a constant magnetic field. In Fig. 6-2, a straight conductor of length l is moving with a velocity v through a uniform magnetic field, the emf in the conductor is:

$$\varepsilon = \int_b^a (\boldsymbol{v} \times \boldsymbol{B}) \cdot d\boldsymbol{l} \tag{6-2}$$

As an example in Fig. 6-2, an emf is produced between two ends of the rod. According to Eq. (6-2), the magnitude of ε is vBl.

Chapter 6 Electromagnetic Inductance

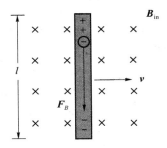

Fig. 6-2 A conducting rod is moving in a uniform magnetic field

3. Induced emf

An induced electric field is suggested by Maxwell in the loop, which exerts a net force on electrons in the loop.

A conducting loop is put in a uniform magnetic field in Fig. 6-3. If B changes in time, an emf is induced, and a current is created in the conducting loop. In this model, the conductor is not moving in the magnetic field, thus there is no magnetic force. Maxwell suggested that there must be an induced electric field created in the conductor as a result of changing magnetic flux. This kind of electric field exists even when no conductor is present. The emf of any closed path can be expressed as:

$$\varepsilon = \oint \boldsymbol{E} \cdot \mathrm{d}\boldsymbol{l} = -\frac{\mathrm{d}\Phi_B}{\mathrm{d}t} = -\int_A \frac{\partial \boldsymbol{B}}{\partial t} \cdot \mathrm{d}\boldsymbol{A} \tag{6-3}$$

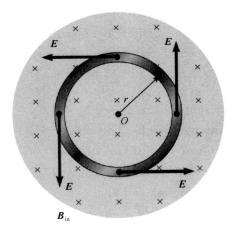

Fig. 6-3 A conducting loop is put in a time-varying magnetic field

where Φ_B is the magnetic flux through the closed loop, and A is the area corresponding to that closed loop. The induced electric field in this equation is a nonconservative field that is generated by a changing magnetic field. It cannot be an electrostatic field because the line integral of $\boldsymbol{E} \cdot \mathrm{d}\boldsymbol{l}$ over a closed loop would not be zero.

4. Self-inductance

An inductor is a circuit element such as solenoid that stores energy in the magnetic field surrounding its current-carrying wires, just as capacitor stores energy in the electric field between its charged plates.

When the current in the coil either increases or decreases with time, the magnetic flux through the coils also changes with time, and induces an emf in the coil. This is a kind of emf called as self-induced emf.

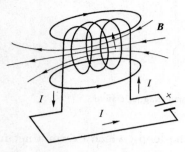

Fig. 6-4   A circuit is consisting of a solenoid

In Fig. 6-4, the emf created by changing the self-current is called self-induced emf. Applying Faraday's law, the induced emf is equal to the negative time rate of change of the magnetic flux. The magnetic flux is proportional to the magnetic field due to the source current, which in turn is proportional to the source current in the circuit. Thus, a self-induced emf is always proportional to the time rate of the change of the source current.

$$\varepsilon_L = -\frac{d(N\Phi_B)}{dt} = -L\frac{dI}{dt} \quad (6\text{-}4)$$

Therefore, we define the self-inductance is:

$$L = \frac{N\Phi_B}{I} \quad (6\text{-}5)$$

Note that, since B is proportional to the current, the self-inductance is independent of I. The self-inductance only depends on the geometry of the device.

5. Mutual inductance

If two coils of wire are placed near each other, a changing current in one will induce an emf in the other. In analogy to the definition of self-inductance, we define the mutual inductance M_{21} of coil 2 with respect to coil 1:

$$M_{21} = \frac{N_2 \Phi_{21}}{I_1} \quad (6\text{-}6)$$

where the Φ_{21} represents the magnetic flux caused by the current in coil 1 and passing through coil 2.

Fig. 6-5 shows a cross-sectional view of two adjacent coils. A current in coil 1 creates a magnetic field and some of magnetic field lines pass through coil 2.

We can also imagine a source current I_2 in coil 2. The definition of M_{21} of coil 1 with respect to coil 2 is:

$$M_{12} = \frac{N_1 \Phi_{12}}{I_2} \quad (6\text{-}7)$$

Basically, we have $M_{21} = M_{12} = M$. The emf in these two coils are $\varepsilon_1 = -M\frac{dI_2}{dt}$ and $\varepsilon_2 = -M\frac{dI_1}{dt}$ respectively.

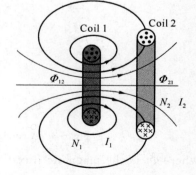

Fig. 6-5   A cross-sectional view of two adjacent coils

6. Energy in a magnetic field

In Fig. 6-6, the battery has negligible internal resistance. This is an RL circuit because the elements connected to the battery are a resistor and an inductor. When the switch is closed, applying Kirchhoff's loop rule to this circuit, we can obtain:

Chapter 6 Electromagnetic Inductance

$$\varepsilon = IR + L\frac{dI}{dt} \qquad (6\text{-}8)$$

The magnetic field is stored in the solenoid.

If we multiply each term by I and rearrange the expression, we have,

$$I\varepsilon = I^2 R + LI\frac{dI}{dt} \qquad (6\text{-}9)$$

and

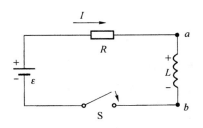

Fig. 6-6 A series RL circuit

$$\int_0^t \varepsilon I\, dt = \int_0^t I^2 R\, dt + \int_0^t LI\frac{dI}{dt} dt \qquad (6\text{-}10)$$

The second term on the right side represents the energy that is delivered from the battery to the inductor and is stored in the magnetic field through the coil. To integrate this expression, we have the total energy:

$$U_B = \int_0^t LI\frac{dI}{dt} dt = \int_0^I LI\, dI = \frac{1}{2}LI^2 \qquad (6\text{-}11)$$

For a solenoid with a cross section of A and length of l, the inductance is given by the following expression,

$$L = \mu_0 n^2 A l \qquad (6\text{-}12)$$

The magnetic field of a solenoid is

$$B = \mu_0 n I \qquad (6\text{-}13)$$

So the energy stored in the solenoid is

$$U_B = \frac{1}{2}LI^2 = \frac{B^2}{2\mu_0} Al \qquad (6\text{-}14)$$

7. Displacement current

Ampère's law can solve the magnetic field with high symmetry. It is only valid if any electric fields present are constant in time. However, if there is a capacitor in the circuit in Fig. 6-7, two surfaces S_1 and S_2 near the plate of a capacitor are bounded by the same path P. The conduction current in the wire passes only through S_1. Therefore, Ampère's law does not satisfy for S_2 surface because no conduction current passes through S_2. Thus, there is a contradictory situation that arises from the discontinuity of the current. Maxwell solved this problem by introducing a fictitious current called displacement current,

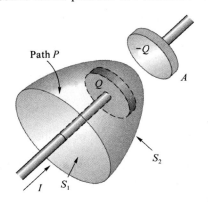

Fig. 6-7 A capacitor is in a circuit

$$I_d = \varepsilon_0 \frac{d\Phi_E}{dt} \qquad (6\text{-}15)$$

where Φ_E is the electric flux.

As the capacitor is being charged, the changing electric field between the plates may be considered to a current that acts as a continuation of the conduction current in the wire. This current (changing electric field) can create magnetic field in the space.

The total current in a circuit is defined by

$$I_{tot} = I + I_d = I + \varepsilon_0 \frac{d\Phi_E}{dt} \tag{6-16}$$

8. Maxwell's equation

J. C. Maxwell is one of the greatest scientists of all time. He developed the electromagnetic theory of light and discovered one of his famous four equations. Individually, these equations are known as Gauss's law, Gauss's law for magnetism, Faraday's law of induction, and Ampère's law with Maxwell's correction.

$$\oint \boldsymbol{E} \cdot d\boldsymbol{A} = \frac{q}{\varepsilon_0}$$

$$\oint \boldsymbol{B} \cdot d\boldsymbol{A} = 0$$

$$\oint \boldsymbol{E} \cdot d\boldsymbol{l} = -\frac{d\Phi_B}{dt} = -\int \frac{\partial \boldsymbol{B}}{\partial t} \cdot d\boldsymbol{A}$$

$$\oint \boldsymbol{B} \cdot d\boldsymbol{l} = \mu_0 I + \varepsilon_0 \mu_0 \frac{d\Phi_E}{dt} = \mu_0 I + \varepsilon_0 \mu_0 \int \frac{\partial \boldsymbol{E}}{\partial t} \cdot d\boldsymbol{A}$$

The last two equations can be combined to obtain a wave equation for both the electric field and the magnetic field. In empty space, the solution to these two equations verifies that the speed of electromagnetic waves is equal to the speed of light. This result led Maxwell to predict that light waves are a form of electromagnetic radiation.

Typical Examples

Problem solving strategy

1. Electromotive force

Emf can be obtained by Faraday's law, or by the definition of motional emf.

(1) Faraday's law of induction. $\varepsilon = -\frac{d\Phi_B}{dt}$. Problem solving procedure is as follows. Usually, B is chosen to be perpendicular to the plane of the loop. The magnetic flux is obtained from $\Phi_B = \int \boldsymbol{B} \cdot d\boldsymbol{A}$. Minus sign reflects Len's law in the problem.

(2) The definition of motional emf. $\varepsilon = \int_b^a (\boldsymbol{v} \times \boldsymbol{B}) \cdot d\boldsymbol{l}$. In this type of problem, the Lorentz force on a segment of the bar of length dr can be obtained. The motional emf induced in this segment is $d\varepsilon = Bvdr$. Generally, every segment of the bar is moving perpendicular to B, and the total emf between the bar can be determined from the definition.

Examples

1. In Fig. 6-8, a conducting bar rotates with a constant angular speed ω about a pivot at one end. A uniform magnetic field is directed perpendicular to the rotation plane. Find the motional emf induced between the ends of the bar.

Chapter 6 Electromagnetic Inductance

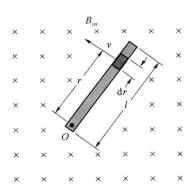

Fig. 6-8 Example 1

Solution: Consider a segment of the bar of length dr with a velocity v. The magnitude of the emf induced in this segment is

$$d\varepsilon = \int (\boldsymbol{v} \times \boldsymbol{B}) \cdot d\boldsymbol{l} = Bv\,dr$$

Summing the emfs of all segment, the total emf between the ends of the bar is:

$$\varepsilon = \int d\varepsilon = \int Bv\,dr = B\omega \int_0^l r\,dr = \frac{1}{2}B\omega l^2$$

2. A short section of wire with length b in Fig. 6-9 is moving with velocity v, parallel to a very long wire carrying a current I. The distance between the near end of the wire section and the long wire is a. Assuming the vertical wire is very long, determine the emf between the ends of the short section.

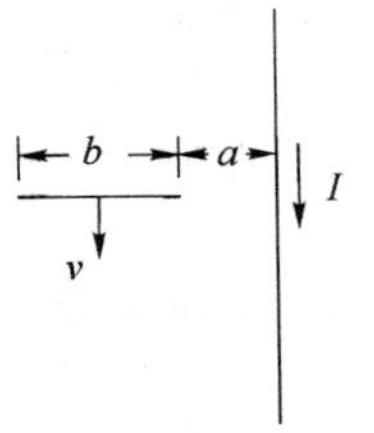

Solution: At a distance r from the long wire, the magnetic field is direction into the paper with magnitude

$$B = \frac{\mu_0 I}{2\pi r}$$

Fig. 6-9 Example 2

Because the field is not constant over the short section, we find the induced emf by integration. The induced emf in a wire of length of dr is,

$$d\varepsilon = Bv\,dr$$

The total emf in the wire is determined by integration,

$$\varepsilon = \int_a^{a+b} Bv\,dr = \frac{\mu_0 Iv}{2\pi} \int_a^b \frac{dr}{r} = \frac{\mu_0 Iv}{2\pi} \ln\left(\frac{a+b}{a}\right)$$

3. A long solenoid of radius R has n turns of wire per unit length and carries a time-varying current as $I = I_m \cos \omega t$. Determine the magnitude of the induced electric field outside the solenoid, a distance $r > R$ from its long central axis.

Solution: Consider an external point and take the path for line integral to be a circle of radius r centered on the solenoid. By symmetry, we see that the magnitude of induced electric field E is constant on this path and that E is tangent to the circle.

Applying the induced electric field,

$$\oint \boldsymbol{E} \cdot d\boldsymbol{l} = -\frac{d(B\pi R^2)}{dt} = -\pi R^2 \frac{dB}{dt}$$

The magnetic field inside a long solenoid is $\mu_0 nI$, we find that

$$\oint \boldsymbol{E} \cdot d\boldsymbol{l} = E(2\pi r) = -\pi R^2 \frac{dB}{dt} = -\pi R^2 \mu_0 nI_{max} \frac{d}{dt}(\cos \omega t)$$

$$E = \frac{\mu_0 nI_{max}\omega R^2}{2r}\sin \omega t$$

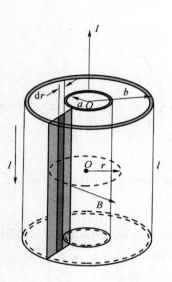

The electric field varies sinusoidally with time and its amplitude changes as $1/r$ outside the solenoid.

4. Coaxial cables are often used to connect electrical devices. Generally, a long coaxial cable is consisting of two thin concentric cylindrical conduction shells of radii a and b and length l as shown in Fig. 6-10. The conducting shells carry the same current I in opposite directions. Imagine that the inner conductor carries current to a device and that the outer one acts as a return path carrying the current back to the source. (1) Calculate the self-inductance L of this cable. (2) Calculate the magnetic energy stored in the magnetic field of the cable.

Solution: Suppose we choose a cross-section with a rectangle shape, the magnetic field is perpendicular to the cross section. Dividing this rectangle into strips of width

Fig. 6-10 Example 4

dr, the magnetic flux passing through each strip is

$$BdA = bldr$$

Hence we find the total flux through the entire cross section by integrating

$$\Phi_B = \int BdA = \int_a^b \frac{\mu_0 I}{2\pi r}ldr = \frac{\mu_0 Il}{2\pi}\ln\left(\frac{b}{a}\right)$$

The self-inductance of the cable is

$$L = \frac{\Phi_B}{I} = \frac{\mu_0 l}{2\pi}\ln\left(\frac{b}{a}\right)$$

The total magnetic energy stored in the cable is

$$U = \frac{1}{2}LI^2 = \frac{\mu_0 lI^2}{4\pi}\ln\left(\frac{b}{a}\right)$$

Questions and Problems

1. Displacement current exists wherever there is _____.
   (A) a magnetic field   (B) an electric field
   (C) a changing electric field (D) a changing magnetic field
2. The magnetic field in a region of space is given by $\boldsymbol{B} = 0.001\, t^2\, \hat{\boldsymbol{i}}$ for $-2\text{ s} \leqslant t \leqslant +2\text{ s}$

(in SI unit). What is the direction of the induced electric field when $t=0$ s? _____.

(A) Parallel to the x axis

(B) Parallel to the y axis

(C) The electric field is in the circles centered on the x axis

(D) There is no induced electric field when $t=0$ s

3. A long, straight wire carries a current I. The magnetic energy density at the point a distance r from the wire is _____.

(A) $\dfrac{1}{2}\mu_0\left(\dfrac{\mu_0 I}{2\pi r}\right)^2$ (B) $\dfrac{1}{2\mu_0}\left(\dfrac{\mu_0 I}{2r}\right)^2$ (C) $\dfrac{1}{2}\mu_0\left(\dfrac{2r}{\mu_0 I}\right)^2$ (D) $\dfrac{1}{2\mu_0}\left(\dfrac{\mu_0 I}{2\pi r}\right)^2$

4. A rectangular loop moves with a velocity perpendicular to a long, straight wire carrying a current I as shown in Fig. 6-11. Which statement is correct for the direction of induced current in the rectangular loop? _____.

(A) There is a counterclockwise and then clockwise current in the loop

(B) There is always a clockwise current in the loop

(C) There is always a counterclockwise current in the loop

(D) There is no induced current in the loop

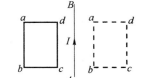

Fig. 6-11 Problem 4

5. A square loop of copper wire in Fig. 6-12 is pulled through a region of magnetic field. Rank in order, from strongest to weakest, the pulling forces that must be applied to keep the loop moving at constant speed. _____.

(A) $F_2=F_4>F_1=F_3$ (B) $F_3>F_2=F_4>F_1$

(C) $F_3>F_4>F_2>F_1$ (D) $F_4>F_2>F_1=F_3$

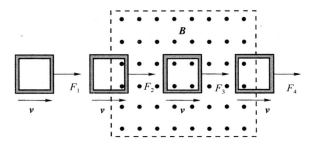

Fig. 6-12 Problem 5

6. A conducting bar in a uniform magnetic field in Fig. 6-13 is rotating about a vertical axis OO' through C. The length of BC is one half of AC. Which of the following about electric potential is true? _____.

(A) $V_A>V_B$ (B) $V_A<V_B$ (C) $V_A=V_B$ (D) $V_C=0$

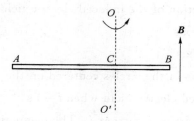

Fig. 6-13   Problem 6

7. A conducting bar of length L rotates with a constant angular speed ω about a pivot at one end as shown in Fig. 6-14. A uniform magnetic field B is directed perpendicular to the plane of rotation. When time is 0, the angle between the conducting bar and the horizontal axis is θ. The motional emf induced between the ends of the bar at time t is _____.

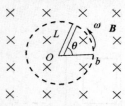

(A) $\omega L^2 B\cos(\omega t+\theta)$

(B) $\frac{1}{2}\omega L^2 B\cos \omega t$

(C) $2\omega L^2 B\cos(\omega t+\theta)$

(D) $\frac{1}{2}\omega L^2 B$

Fig. 6-14   Problem 7

8. Consider the two closely wound coils of wire with area of A and $2A$ respectively as shown in Fig. 6-15. The magnetic flux caused by the current I in coil 1 and passing through coil 2 is represented by Φ_{21}, while the magnetic flux caused by the current I in coil 2 and passing through coil 2 is represented by Φ_{12}. Which one is correct for the relationship of magnetic flux? _____.

(A) $\Phi_{21}=2\Phi_{12}$ (B) $\Phi_{21}>\Phi_{12}$

(C) $\Phi_{21}=\Phi_{12}$ (D) $\Phi_{21}=\frac{1}{2}\Phi_{12}$

Fig. 6-15   Problem 8

9. The capacitor in Fig. 6-16 is connected to a battery and is being charged slowly. Consider these two line integral of magnetic strength, the correct one is _____.

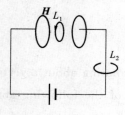

(A) $\oint_{L_1} \boldsymbol{H} \cdot \mathrm{d}\boldsymbol{l}' > \oint_{L_2} \boldsymbol{H} \cdot \mathrm{d}\boldsymbol{l}'$

(B) $\oint_{L_1} \boldsymbol{H} \cdot \mathrm{d}\boldsymbol{l}' = \oint_{L_2} \boldsymbol{H} \cdot \mathrm{d}\boldsymbol{l}'$

(C) $\oint_{L_1} \boldsymbol{H} \cdot \mathrm{d}\boldsymbol{l}' < \oint_{L_2} \boldsymbol{H} \cdot \mathrm{d}\boldsymbol{l}'$

(D) $\oint_{L_1} \boldsymbol{H} \cdot \mathrm{d}\boldsymbol{l}' = 0$

Fig. 6-16   Problem 9

Chapter 6　Electromagnetic Inductance

10. Consider a circuit consisting of a battery connected to two conducting rod. The current in the circuit is I. What is the magnetic energy density at point P in the middle, and with equal distance a from each rod in Fig. 6-17? _____.

(A) $\dfrac{1}{\mu_0}\left(\dfrac{\mu_0 I}{2\pi a}\right)^2$ 　　　(B) $\dfrac{1}{2\mu_0}\left(\dfrac{\mu_0 I}{2\pi a}\right)^2$

(C) $\dfrac{1}{2\mu_0}\left(\dfrac{\mu_0 I}{\pi a}\right)^2$ 　　　(D) 0

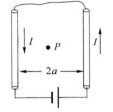

Fig. 6-17　Problem 10

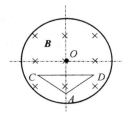

Fig. 6-18　Problem 11

11. A magnetic field $\boldsymbol{B}$ is confined to columnar space in Fig. 6-18. Suppose the magnetic field is changing with time according to $\dfrac{d\boldsymbol{B}}{dt}>0$. Consider an isosceles triangle-shaped conducting coil in this magnetic field, the emf induced in CD is ε_1, the emf induced in CAD is ε_2, and the emf induced in triangle-shaped coil $ACDA$ is ε. Which of the following conclusion is correct? _____.

(A) $\varepsilon_1=-\varepsilon_2$,　$\varepsilon=\varepsilon_1+\varepsilon_2=0$
(B) $\varepsilon_1>0,\varepsilon_2<0$,　$\varepsilon=\varepsilon_1+\varepsilon_2>0$
(C) $\varepsilon_1>0,\varepsilon_2>0$,　$\varepsilon=\varepsilon_1-\varepsilon_2<0$
(D) $\varepsilon_1>0,\varepsilon_2>0$,　$\varepsilon=\varepsilon_2-\varepsilon_1>0$

12. The toroid in Fig. 6-19 consists of N turns and has a rectangular cross section. Its inner and outer radii are a and b, respectively. What is the inductance of the toroid? _____.

(A) $\dfrac{\mu_0 N^2 (b-a)h}{2\pi a}$ 　　　(B) $\dfrac{\mu_0 N^2 h}{2\pi}\ln\dfrac{b}{a}$

(C) $\dfrac{\mu_0 N^2 (b-a)h}{2\pi b}$ 　　　(D) $\dfrac{\mu_0 N^2 (b-a)h}{2\pi (b+a)}$

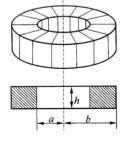

Fig. 6-19　Problem 12

13. Consider two iron loop and cooper loop with same size. When magnetic flux in these loops are changing in the same ratio, the correct statement is that _____.

(A) both loops have different emf and induced current
(B) both loops have same emf and induced current
(C) both loops have different emf and same induced current
(D) both loops have same emf and different induced current

14. The circuit in Fig. 6-20 consists of a resistor, an inductor, a capacitor, and an ideal battery with no internal resistance. After the switch is closed, and the current is in steady-state. Once the switch is open, the maximum electric potential difference between the ends

of capacitor is _____ .

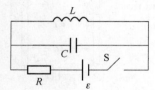

Fig. 6-20 Problem 14

(A) $\dfrac{\varepsilon}{R}\sqrt{LC}$ (B) $\dfrac{\varepsilon}{R}\sqrt{\dfrac{1}{LC}}$

(C) $\dfrac{\varepsilon}{R}\sqrt{\dfrac{C}{L}}$ (D) $\dfrac{\varepsilon}{R}\sqrt{\dfrac{L}{C}}$

15. Consider two uniformly wound solenoids have same length and turns. Their radii are r_1 and r_2, respectively. There are full of magnetic substance inside, and the magnetic permeability are μ_1 and μ_2 respectively. Suppose $r_1 : r_2 = 1 : 2$ and $\mu_1 : \mu_2 = 2 : 1$. Once we connect the solenoids to closed circuits, the ratios of inductance and magnetic energy are _____ .

(A) $L_1 : L_2 = 1 : 1, U_{m1} : U_{m2} = 1 : 1$
(B) $L_1 : L_2 = 1 : 2, U_{m1} : U_{m2} = 1 : 2$
(C) $L_1 : L_2 = 1 : 2, U_{m1} : U_{m2} = 1 : 1$
(D) $L_1 : L_2 = 2 : 1, U_{m1} : U_{m2} = 2 : 1$

16. The electric field in four identical capacitors is shown in Fig. 6-21 as a function of time. Rank in order, from largest to smallest, the magnetic field strength at the outer edge of the capacitor at time T. _____ .

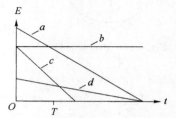

Fig. 6-21 Problem 16

(A) $B_a = B_b > B_c = B_d$ (B) $B_a > B_b > B_c > B_d$
(C) $B_a = B_b > B_c > B_d$ (D) $B_c > B_a > B_d > B_b$

17. Experimenter A creates a magnetic field in the laboratory. Experimenter B moves relative to A. Experimenter B sees _____ .

(A) a magnetic field of different strength
(B) a magnetic field pointing the opposite direction
(C) just an electric field
(D) both a magnetic and an electric field

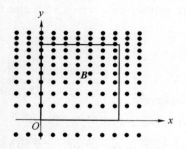

Fig. 6-22 Problem 19

18. A cylindrical region of radius $R = 3.0$ cm contains a uniform magnetic field parallel to its axis. If the field is changing at the rate 0.60 T/s, the electric field induced at a point $R/2$ from the cylinder axis is _____ .

19. The square shaped wire in Fig. 6-22 has sides of length 2.0 cm. A magnetic field points out of the page; its magnitude is given by $B = 4t^2 y$ (SI). The magnitude of emf around the square wire at $t = 2.5$ s is _____ V. The direction of emf is _____

(clockwise or counterclockwise).

20. The long, straight wire in Fig. 6-23 carries a current I to the right that decreases linearly with time. The direction of the induced current in the circular wire loop is _____ (clockwise or counterclockwise).

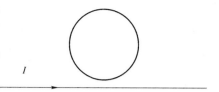

Fig. 6-23 Problem 20

21. Write the correct forms of four Maxwell's equations in the following blanks.

Gauss's law for electricity _____.

Gauss's law for magnetism _____.

Faraday's law of induction _____.

Generalized Ampère's law _____.

22. Fig. 6-24 shows a rod of length L caused to move at constant speed v along horizontal conducting rails. The magnetic field in which the rod moves is provided by a current i in a long wire parallel to the rails. Assume $v=5.00$ m/s, $a=10.0$ mm, $L=9.0$ cm, $i=100$ A, and the total resistance of rod and rails $R=0.400$ Ω. (1) Calculate the emf induced in the rod. (2) What is the current in the conducting loop? ($\ln 10 = 2.3$)

23. A long straight wire and a conducting rod, carrying separately current I_1 and I_2, are placed on a plane in Fig. 6-25. The rod at an angle θ with the horizontal line of the wire. Find the resultant force acting on the rod.

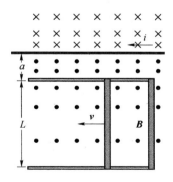

Fig. 6-24 Problem 22

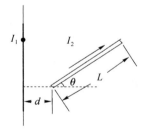

Fig. 6-25 Problem 23

24. Two straight, conducting rails form an angle θ where their ends are jointed, as shown in Fig. 6-26. A conducting bar MN in contact with the rails (and forming a triangle loop with the rails) slides to the right with a constant speed $v (v \perp MN)$. The initial position of bar $x=0$ at time $t=0$. A magnetic field $\boldsymbol{B}$ points out of the page. Find the induced emf in the loop as a function of time, if (1) the magnetic field is uniform and has value B, or (2) a non-uniform magnetic field follows the function $B = K\xi \cos \omega t$, where ξ is the horizontal coordinate.

25. A rectangular loop of wire with length a, width b and resistance R in Fig. 6-27 is placed near an infinitely long wire carrying current i. The distance between the wire and the nearest edge of the loop is c. Find (1) the magnitude of the magnetic flux through the loop and (2) the induced current i_{ind} in the loop as it moves away from the long wire with speed v.

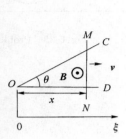

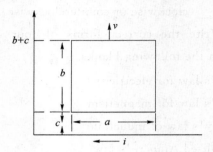

Fig. 6-26 Problem 24   Fig. 6-27 Problem 25

26. Estimate the mutual inductance M between two small loops of radii r_1 and r_2, which are separated by a distance l that is large compared to r_1 and r_2, as shown in Fig. 6-28. Give M as a function of θ, the angle between the plans of the two coils. Assume the line joining their centers is perpendicular to the plane of coil 1.

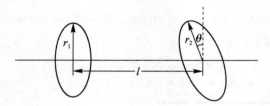

Fig. 6-28 Problem 26

27. A cylindrical conductor of radius R and conductivity σ in Fig. 6-29 carries a steady current I distributed uniformly across its cross-section. (1) Determine E inside the conductor. (2) Determine B just outside the conductor.

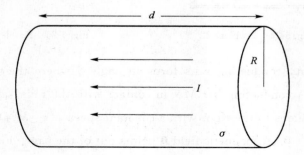

Fig. 6-29 Problem 27

Chapter 7 Kinetic Theory of Gases

Review of the Contents

1. Kinetic theory of gases

Kinetic theory is the theory that gases are made up of a large number of small particles (atoms or molecules), all of which are in constant, random motion. The rapidly moving particles constantly collide with each other and with the walls of the container. Kinetic theory explains macroscopic properties of gases, such as pressure, temperature, or volume, by considering their molecular composition and motion. The essence of the theory is that pressure is due to collisions between molecules moving at different velocities.

2. Ideal gases

An ideal gas is one that obeys the ideal gas law. At low to moderate pressures, and at temperatures not too low, these gases can be considered as ideal gases: air, nitrogen, oxygen, hydrogen etc. A real gas behaves like an ideal gas when its atoms or molecules are so far apart that they do not interact with each other.

Generally, the ideal gases can be defined by two following ways.

(1) Macroscopic description of ideal gases

For a gas, it is useful to know the quantities volume V, pressure p and temperature T. In general, the equation that interrelates these quantities, called the equation of state.

$$pV = \frac{m}{M}RT = nRT \tag{7-1}$$

where $R = 8.314$ J/(mol · K) is the universal gas constant. If the volume contains m kilograms of gas that has a molar mass M, then $n = m/M$.

Sometimes, the ideal gas law is expressed in terms of the total number of molecules N. We can get,

$$pV = nRT = \frac{N}{N_A}RT = NkT \tag{7-2}$$

where k is Boltzmann's constant.

(2) Microscopic description of ideal gases

In microscopic view, an ideal gas model is valid based on these assumptions:
- The number of molecules is large in the gas and the average separation between them is large enough compared with its size.
- Newton's laws of motion are valid for each molecule.
- The molecules interact only by short-range forces during elastic collisions.
- All molecules are identical.

3. Pressure of ideal gas

The pressure exerted by the gas on the walls of its container is due to the collision of the molecules with the walls, as shown in Fig. 7-1. The pressure exerted by n moles of an ideal gas, in terms of the speed of its molecules, is

$$p = \frac{1}{3}nm\overline{v^2} \quad \text{or} \quad p = \frac{2}{3}n\overline{\varepsilon_t} \tag{7-3}$$

where the square root of $\overline{v^2}$ is called the root-mean-square speed of the molecules and $\overline{\varepsilon_t}$ is the average translational kinetic energy of each molecule.

$$v_{rms} = \sqrt{\overline{v^2}} = \sqrt{\frac{3RT}{M}} \tag{7-4}$$

where M is the molar mass in kilograms per mole. In this expression, we can see that, at a given temperature, lighter molecules move faster.

The average translational kinetic energy of each molecule is

$$\overline{\varepsilon_t} = \frac{1}{2}m\overline{v^2} = \frac{3}{2}kT \tag{7-5}$$

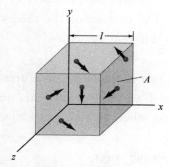 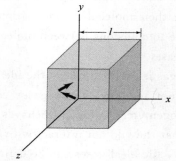

(a) Molecules of an ideal gas moving about in a rectangular container

(b) Arrows indicate the momentum of one molecule as it rebounds from the end wall

Fig. 7-1 The connection between the pressure exerted by the gas on the walls and the action of bumping into the walls by the molecules of the gas

It is obvious that the average translational kinetic energy of molecules in an ideal gas is directly proportional to the absolute temperature.

4. The equipartition of energy

The theorem of equipartition of energy states that molecules in thermal equilibrium have the same average energy associated with each independent degree of freedom (DOF) of their motion.

Every degree of freedom of a molecule has associated with energy $\frac{1}{2}kT$ per molecule. If i is the number of degrees of freedom, then the average kinetic energy is

$$\overline{\varepsilon_k} = \frac{i}{2}kT \tag{7-6}$$

For monatomic gases, internal energy is stored only in translational motion of the atoms. It moves as point-like body and can move in three independent directions.

In Fig. 7-2, for diatomic and triatomic molecules, rotational motion is significant. Generally, it will add another two more DOFs. In this case, both translational and rotational motion can contribute to the kinetic energy. Table 7-1 gives the average energy associated with different types of gases.

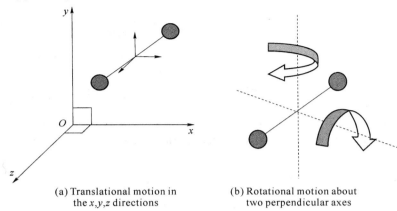

(a) Translational motion in the x,y,z directions

(b) Rotational motion about two perpendicular axes

Fig. 7-2 A model of diatomic molecule

Table 7-1 Average kinetic energy associated with different types of gases

Type	Examples	DOF	Average kinetic energy
Monatomic gas	He, Ne, Ar	3	$1.5\,kT$
Diatomic gas	O_2, H_2, CO	5	$2.5\,kT$
Triatomic gas	H_2O, CO_2, NH_3	6	$3\,kT$

5. Distribution of molecular speeds

In 1860 Maxwell gave an expression about the distribution of molecular speeds in a very definite manner. About 60 years later, his prediction was confirmed by experiments.

The fundamental expression that describes the distribution of speeds of N gas molecules is

$$f(v) = 4\pi \left(\frac{m}{2\pi kT}\right)^{\frac{3}{2}} v^2 e^{-\frac{mv^2}{2kT}} \quad (7-7)$$

where m is the mass of a gas molecule, k is Boltzmann's constant, and T is absolute temperature. This expression is plotted in Fig. 7-3.

In Fig. 7-3, $f(v)\,\mathrm{d}v$ gives the fraction of molecules with speeds in the interval $\mathrm{d}v$ centered on speed v. The number of molecules having speeds in the range v_1 to v_2 is equal to the area of the shaded shape. v_p is the most probable speed where $f(v)$ has its maximum value. The probability of this speed occurs more than any others. We can see that the function $f(v)$ approaches zero as v approaches infinite.

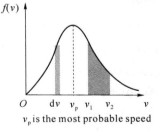

v_p is the most probable speed

Fig. 7-3 Distribution of molecular speeds at some temperature

Typical Examples

Problem solving strategy

1. The Maxwell speed distribution function

In this chapter, a typical problem is that to apply Maxwell speed distribution function.

So, it is important to learn the physical meaning of these expressions.

(1) $f(v)dv = \dfrac{dN_v}{N}$. The proportion of molecules that have speed between v and $v+dv$.

(2) $\int_{v_1}^{v_2} f(v)dv = \dfrac{\Delta N}{N}$. The proportion of molecules that have speed between v_1 and v_2.

(3) $\int_0^\infty f(v)dv = 1$. The unitary condition of the speed distribution function.

(4) $Nf(v)dv = dN_v$. The number of molecules that have speed between v and $v+dv$.

2. Three measures of the speeds distribution of the molecules

(1) Most probable speed v_p

$$v_p = \sqrt{\dfrac{2RT}{M}} = 1.41\sqrt{\dfrac{kT}{m}} \tag{7-8}$$

(2) Average speed $\bar{v}$

$$\bar{v} = \int_0^\infty v f(v)dv = \sqrt{\dfrac{8RT}{\pi M}} = 1.60\sqrt{\dfrac{kT}{m}} \tag{7-9}$$

(3) The root-mean-square-speed $\sqrt{\overline{v^2}}$

$$\sqrt{\overline{v^2}} = \int_0^\infty v^2 f(v)dv = \sqrt{\dfrac{3RT}{M}} = 1.73\sqrt{\dfrac{kT}{m}} \tag{7-10}$$

These equations can be used in different aspects. We can see that at some temperature, $\sqrt{\overline{v^2}} > \bar{v} > v_p$.

Examples

1. A container is filled with oxygen gas maintained at room temperature (300 K). What fraction of the molecules have speeds in the interval 599 to 601 m/s? The molar mass M of oxygen is 0.032,0 kg/mol.

Solution:

The speeds of the molecules are distributed over a wide range of values, with the distribution function

$$f(v) = 4\pi \left(\dfrac{m}{2\pi kT}\right)^{\frac{3}{2}} v^2 e^{-\frac{mv^2}{2kT}}$$

The fraction of the molecules with speeds in a differential interval dv is $f(v)dv$. The interval $\Delta v = 2$ m/s here is small compared to the speed $v = 600$ m/s on which it is centered. Thus we can avoid the integration by approximating the fraction as

$$\text{frac.} = f(v)\Delta v = 4\pi \left(\dfrac{m}{2\pi kT}\right)^{\frac{3}{2}} v^2 e^{-\frac{mv^2}{2kT}} \Delta v = 2.62 \times 10^{-3}$$

At room temperature, 0.262% of the oxygen molecules will have speeds in the narrow range between 599 and 601 m/s.

2. Starting from the Maxwell distribution of speeds $f(v) = 4\pi \left(\dfrac{m}{2\pi kT}\right)^{\frac{3}{2}} v^2 e^{-\frac{mv^2}{2kT}}$, shows

Chapter 7 Kinetic Theory of Gases

(1) $\int_0^\infty f(v)\,dv = 1$ and (2) $\int_0^\infty v^2 f(v)\,dv = 3\dfrac{kT}{m}$.

Solution:

(1) The total number of molecules is represented by the area under the Maxwell distribution, which we find by integration

$$\int_0^\infty f(v)\,dv = \int_0^\infty 4\pi \left(\frac{m}{2\pi kT}\right)^{3/2} v^2 e^{-mv^2/2kT}\,dv$$

This can be simplified by a change in variable:

$$u = \left(\frac{m}{2kT}\right)^{1/2} v; \quad du = \left(\frac{m}{2kT}\right)^{1/2} v\,dv$$

The integral becomes

$$\int_0^\infty f(v)\,dv = \int_0^\infty 4\pi \left(\frac{1}{\pi}\right)^{3/2} u^2 e^{-u^2}\,du = 4\pi \left(\frac{1}{\pi}\right)^{3/2}\left(\frac{\pi^{1/2}}{4}\right) = 1$$

where we have used the result from integral tables.

(2) We make the same change in variable for the integral to find the rms speed:

$$\int_1^\infty v^2 f(v)\,dv = \int_0^\infty 4\pi \left(\frac{m}{2\pi kT}\right)^{3/2} v^4 e^{-mv^2/2kT}\,dv$$

$$= 4\pi \left(\frac{1}{\pi}\right)^{3/2}\left(\frac{2kT}{m}\right)\int_0^\infty u^4 e^{-u^2}\,du = 4\pi\left(\frac{1}{\pi}\right)^{3/2}\left(\frac{2kT}{m}\right)\left(\frac{3\pi^{1/2}}{8}\right) = \frac{3kT}{m}$$

Questions and Problems

1. For two objects have the same temperature, they must _____.
   (A) be in thermal equilibrium
   (B) be in thermal contact with each other
   (C) have the same specific heat
   (D) have the same relative "hotness" or "coolness" when touched

2. A thermometer is made by attaching a capillary tube (diameter 0.1 mm) to a 1 cm³ reservoir of mercury. What length in cm is needed for expansion for a change in temperature of 100°C? _____. The volume expansion coefficient of mercury is $1.82 \times 10^{-4} \,°C^{-1}$.
   (A) 60 (B) 230 (C) 180 (D) 360

3. When comparing the root-mean-square speed $\sqrt{\overline{v^2}}$, the average speed $\overline{v}$, and the most probable speed v_p, of molecules in a gas at temperature T, we find that _____.
   (A) $\sqrt{\overline{v^2}} < \overline{v}$ (B) $\sqrt{\overline{v^2}} = \overline{v}$ (C) $\sqrt{\overline{v^2}} > \overline{v}$ (D) $v_p > \overline{v}$

4. The meaning of the Maxwell distribution function, $f(v)\,dv$ for N molecules in thermal equilibrium follows from the definition of the number of molecules with speeds between v and $v+dv$ as _____.
   (A) $N\,dv$ (B) $f(v)\,dv$ (C) $\dfrac{f(v)}{N}\,dv$ (D) $f(v)N\,dv$

5. A 0.050 kg lump of solid CO_2 is sealed in a 5.0 L evacuated container. The solid CO_2 is heated to 300 K. The pressure in the container is _____ due to the CO_2 gas(One mole of

CO_2 has a mass of 44 g).

6. An H_2 molecule at 293 K has a molar mass of 2.02×10^{-3} kg/mol. If $R = 8.31$ J/(mol·K), its rms speed is _____ m/s.

7. A pressure of 1.33 kPa is measured using a constant-volume gas thermometer at a temperature of 50℃. The pressure is _____ kPa at the zero-point temperature.

8. An isolated container holds 4 moles of hydrogen ($M = 2$ g/mol) and 2 moles of helium ($M = 4$ g/mol) at 400 K. The ratio of the total kinetic energy of the hydrogen molecules to the total kinetic energy of the helium molecules is _____.

9. The speed distribution function of N molecules is shown in Fig. 7-4, when v is larger than $2v_0$, the speed distribution approaches zero. Find (1) constant a; (2) the number of molecules located at $0 \sim v_0$ and $v_0 \sim 2v_0$ respectively; (3) the average speed of molecules.

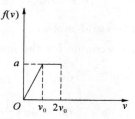

Fig. 7-4   Problem 9

Chapter 8 The First Law of Thermodynamics

Review of the Contents

1. Work in thermodynamic processes

A thermodynamics process is defined as energetic evolution of a thermodynamic system proceeding from an initial state to a final state. In this process, work performed by a system is the quantity of energy transferred by the system to another due to changes in the external parameters of the system. In Fig. 8-1, a gas may exchange energy with its surroundings through work. The amount of work W done by a gas as it expands or contracts from an initial volume V_i to a final V_f is given by

$$W = \int_{V_i}^{V_f} p \, dV \tag{8-1}$$

which is equal to the area under the pV curve.

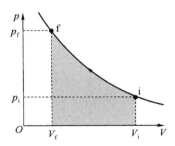

Fig. 8-1 A gas is compressed quasi-statically from state i to state f

2. Heat and internal energy

There is a major distinction between internal energy and heat. Heat marks the transfer of energy between a system and its environment due to a temperature difference. Internal energy is the energy associated with the microscopic components of a system. The internal energy includes kinetic and potential energy of the particles and any potential energy bonding the particles together.

For an ideal gas, the internal energy is expressed as

$$U = N\bar{\varepsilon} = N \frac{i}{2} kT = \frac{N}{N_A} \cdot \frac{i}{2} N_A kT = n \frac{i}{2} RT \tag{8-2}$$

According to the equipartition of energy, the internal energy of different type of gases is listed in Table 8-1.

Table 8-1 The internal energy of different gases

Type	Examples	DOF	Internal energy
Monatomic gas	He, Ne, Ar	3	$(3/2)nRT$
Diatomic gas	O_2, H_2, CO	5(rigid)	$(5/2)nRT$
Triatomic gas	H_2O, CO_2, NH_3	6(rigid)	$(6/2)nRT$

3. The first law of thermodynamics

The first law of thermodynamics states that energy can be transformed but cannot be created or destroyed. The content of the first law is that, "The increase in the internal energy of a system is equal to the amount of energy added by heating the system, minus the amount lost as a result of the work done by the system on its surroundings."

The expression of the first law of thermodynamics is

$$\Delta U = Q - W \tag{8-3}$$

When a system undergoes an infinitesimal change, the expression of the first law is written as

$$dQ = dU + dW \tag{8-4}$$

In the first law of thermodynamics, heat Q is positive if the system absorbs heat, and negative if the system loses heat; W is positive if the system expands against some external force exerted by the surroundings, and negative if the system contracts because of some external force. Both Q and W are path dependent.

ΔU means the change of internal energy of the system. It is a path independent quantity.

4. Heat capacity and specific heat

The heat capacity C of a particular sample of a substance is defined as the amount of energy needed to raise the temperature of that sample by 1℃. The expression of heat capacity is

$$C = \frac{\Delta Q}{\Delta T} = \frac{dQ}{dt} \tag{8-5}$$

If energy Q produces a change from T_1 to T_2, heat Q absorbed in this process is expressed as

$$Q = C(T_2 - T_1) \tag{8-6}$$

The specific heat c of a substance is the heat capacity per unit mass. Then the specific heat of the substance is

$$Q = cm(T_2 - T_1) \tag{8-7}$$

The molar specific heat of a substance is the heat capacity per mole,

$$C_{\text{mol}} = \frac{C}{n} \tag{8-8}$$

The amount of heat energy required to raise the temperature of one mole of a gas by one Kelvin at constant pressure/volume is called molar specific heat at constant pressure/volume.

Chapter 8 The First Law of Thermodynamics

In the process at constant volume, $dV=0$, the change of internal energy is

$$dU = dQ - dW = dQ - pdV = dQ \qquad (8\text{-}9)$$

So, we can get the expression for the $C_{V,m}$,

$$C_{V,m} = \frac{1}{n}\left(\frac{dQ}{dT}\right)_V = \frac{1}{n}\left(\frac{dU}{dT}\right) = \frac{i}{2}R \qquad (8\text{-}10)$$

The molar specific heat at constant pressure is

$$C_{p,m} = \frac{1}{n}\frac{dQ}{dT} = \frac{1}{n}\frac{dU}{dT} + \frac{pdV}{ndT} = C_{V,m} + \frac{pdV}{ndT} \qquad (8\text{-}11)$$

For an ideal gas, applying the equation of state, $pdV = nRdT$, we can obtain the Mayer formula,

$$C_{p,m} = C_{V,m} + R \qquad (8\text{-}12)$$

We can define the specific heat ratio, and the expression is,

$$\gamma = \frac{C_{p,m}}{C_{V,m}} = \frac{i+2}{i} \qquad (8\text{-}13)$$

Table 8-2 The specific heat of different type of gases

Type	Examples	DOF	Internal energy	$C_{V,m}$	$C_{p,m}$	γ
Monatomic gas	He, Ne, Ar	3	$(3/2)nRT$	$(3/2)R$	$(5/2)R$	5/3
Diatomic gas	O_2, H_2, CO	5	$(5/2)nRT$	$(5/2)R$	$(7/2)R$	7/5
Triatomic gas	H_2O, CO_2, NH_3	6	$(6/2)nRT$	$(6/2)R$	$(8/2)R$	4/3

5. Heat engine and cycle process

Heat engine is defined as a device that converts heat energy into mechanical energy or more exactly a system which operates continuously and only heat and work may pass across its boundaries. The operation of a heat engine can best be represented by a thermodynamic cycle.

(1) Cycle processes

In a cycle process of Fig. 8-2, the working substance absorbs energy by heat from a high-temperature energy reservoir, work is done by the engine, and energy is expelled by heat to a lower-temperature reservoir.

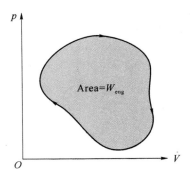

The internal energy in a cycle process doesn't change, thus the net work done in this process is equal to the net energy Q transferred to it.

$$Q = W \qquad (8\text{-}14)$$

Fig. 8-2 A p-V diagram for an arbitrary cyclic process taking place in an engine

The net work done by the engine equals the area enclosed by the curve.

(2) Heat engines

An engine shown in Fig. 8-3 is a device that, operating in a cycle, extracts energy $|Q_h|$ as heat from a high-temperature reservoir and does a certain amount of work W_{eng}.

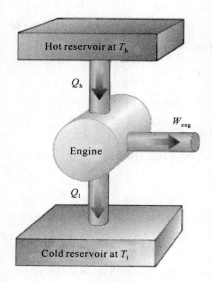

Fig. 8-3　A schematic diagram of energy transfer for a heat engine

The thermal efficiency ε of any heat engine is defined as

$$\varepsilon = \frac{W_{eng}}{|Q_h|} = \frac{|Q_h| - |Q_l|}{|Q_h|} = 1 - \frac{|Q_l|}{|Q_h|} \tag{8-15}$$

In this equation, Q_h represents the energy from hot reservoir. And Q_l means the energy to the cold reservoir.

(3) Heat pumps

A heat pump shown in Fig. 8-4 is a device that, operating in a cycle, has work W done on it as it extracts energy $|Q_l|$ as heat from a low-temperature reservoir and expels energy to a high-temperature reservoir. The coefficient of performance η of a refrigerator is defined as

$$\eta = \frac{|Q_l|}{W} = \frac{|Q_l|}{|Q_h| - |Q_l|} \tag{8-16}$$

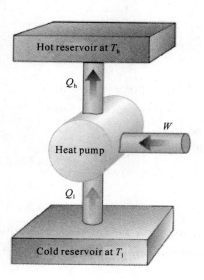

Fig. 8-4　A schematic diagram of a heat pump

Chapter 8   The First Law of Thermodynamics

(4) Carnot cycle

Carnot cycle is an ideal, reversible cycle as shown in Fig. 8-5. It is operating between two energy reservoirs with four quasi-static processes.

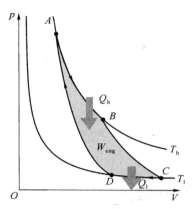

Fig. 8-5   A p-V diagram for the Carnot cycle

Fig. 8-5 shows that the net work equals the net energy transferred into the Carnot engine. Process $A \rightarrow B$ and $C \rightarrow D$ are isothermal expansion and isothermal compression. Process $B \rightarrow C$ and $D \rightarrow A$ are adiabatic processes. The arrows on the piston indicate the direction of its motion during each process.

The efficiency of Carnot cycle is

$$\varepsilon_C = 1 - \frac{|Q_l|}{|Q_h|} = 1 - \frac{T_l}{T_h} \tag{8-17}$$

where T_h and T_l are the temperatures of the high and low temperature reservoirs, respectively.

Typical Examples

Problem solving strategy

Four typical thermodynamics processes

(1) An adiabatic process ($Q=0$)

An adiabatic process is a thermodynamic process in which no heat enters or leaves a system during expansion or compression of the system.

The mathematical equation for an ideal gas undergoing a reversible adiabatic process is

$$pV^{\gamma} = \text{constant} \tag{8-18}$$

Using the ideal gas equation $pV = nRT$, we can write this expression in two other useful forms,

$$TV^{\gamma-1} = \text{constant} \tag{8-19}$$

$$p^{1-\gamma}T^{\gamma} = \text{constant} \tag{8-20}$$

(2) Isochoric process

An isochoric process, also called a constant-volume process, is a thermodynamic process during which the volume of the closed system undergoing such process remains constant.

In this process, there is no pressure-volume work. We can get

$$W = \int p dV = 0 \tag{8-21}$$

By applying the first law of thermodynamics, we can deduce that ΔU the change in the system's internal energy, is

$$\Delta U = Q = n C_{V,m} \Delta T \tag{8-22}$$

(3) Isobaric process

An isobaric process is a thermodynamic process during which the pressure of the system remains constant.

The net work done on the system can be obtained by

$$W = \int p dV = p \Delta V \tag{8-23}$$

The internal energy in the process can be expressed as

$$\Delta U = n C_{V,m} \Delta T \tag{8-24}$$

So we can get the heat change in the process. The formula is

$$Q = W + \Delta U$$
$$= p \Delta V + n C_{V,m} \Delta T = n C_{p,m} \Delta T \tag{8-25}$$

(4) Isothermal process

A thermodynamic process in which the temperature of the system remains constant during the supply of heat is called an isothermal process.

According to the first law of thermodynamics, there is no change in the internal energy of the system, we can get

$$Q = W + \Delta U = W \tag{8-26}$$

Examples

1. A bubble of 5.00 mol of helium is submerged at a certain depth in liquid water when the water (and thus the helium) undergoes a temperature increase ΔT of 20.0 ℃ at constant pressure. As a result, the bubble expands. The helium is monatomic and ideal.

(1) How much energy is added as heat to the helium during the temperature increase?

(2) What is the change ΔU in the internal energy of the helium during the temperature increase?

(3) How much work W done by the helium as it expands against the pressure of the surrounding water during the temperature increase?

Solution:

(1) Because the pressure p is keeping constant during the addition of energy, the heat energy can be expressed as:

$$Q = n C_{p,m} \Delta T$$

For monatomic gas $C_{p,m} = 5/2R$, thus

Chapter 8 The First Law of Thermodynamics

$$Q = nC_{p,m}\Delta T = 5.0 \text{ mol} \times 2.5 \times 8.31 \text{ J/(mol·K)} \times 20 \text{ K} = 2,077.5 \text{ J}$$

(2) The change of internal energy ΔU is:

$$\Delta U = nC_{V,m}\Delta T = n3/2R\Delta T = 5.0 \text{ mol} \times 1.5 \times 8.31 \text{ J/(mol·K)} \times 20 \text{ K} = 1,246.5 \text{ J}$$

(3) The work done by gas expanding against the pressure from its environment is given by $W = \int_{V_1}^{V_2} p dV$. When the pressure of ideal gas system is constant, we can get

$$W = p\Delta V = nR\Delta T$$

$$W = nR\Delta T = 5.0 \text{ mol} \times 8.31 \text{ J/(mol·K)} \times 20 \text{ K} = 831 \text{ J}$$

2. 1 mol of oxygen expands isothermally (at 310 K) from an initial volume of 12 L to a final volume of 19 L.

(1) What would be the final temperature if the gas had expanded adiabatically to this same final volume?

(2) What would be the final temperature and pressure if, instead, the gas had expanded freely to the final volume, from an initial pressure of 2.0 Pa?

Solution:

(1) We can relate the initial and final temperatures and volumes with

$$T_i V_i^{\gamma-1} = T_f V_f^{\gamma-1}$$

Because the molecules are diatomic and the specific heats ratio is

$$\gamma = \frac{C_{p,m}}{C_{V,m}} = \frac{7R/2}{5R/2} = 1.40$$

Then

$$T_f = \frac{T_i V_i^{\gamma-1}}{V_f^{\gamma-1}} = \frac{310 \text{ K} \times (12 \text{ L})^{1.40-1}}{(19 \text{ L})^{1.40-1}} = 258 \text{ K}$$

(2) The temperature does not change in a free expansion:

$$T_f = T_i = 310 \text{ K}$$

We find the new pressure using $p_i V_i = p_f V_f$, which gives us

$$p_f = p_i \frac{V_i}{V_f} = 2.0 \text{ Pa} \times \frac{12 \text{ L}}{19 \text{ L}} = 1.3 \text{ Pa}$$

3. 1 mol O_2 is taken a cycle shown in Fig. 8-6. Path ab is an isothermal process, path bc is an isobaric process, path ca is an isochoric process. Calculate the thermal efficiency of this cycle.

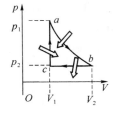

Fig. 8-6 Example 3

Solution:

In the figure, we can get the total input heat of one cycle,

$$Q_1 = Q_{ab} + Q_{ca}$$

$$= \frac{m}{M} RT_A \ln \frac{V_2}{V_1} + \frac{m}{M} C_{V,m}(T_a - T_c)$$

$$= p_1 V_1 \ln \frac{V_2}{V_1} + \frac{5}{2}(p_1 V_1 - p_2 V_1)$$

The total output heat of one cycle is,

$$Q_2 = Q_{bc} = \left| \frac{m}{M} C_{p,m}(T_c - T_b) \right|$$

$$= \frac{7}{2}(p_2 V_2 - p_2 V_1)$$

Thermal efficiency is

$$\varepsilon = 1 - \frac{|Q_2|}{Q_1} = 1 - \frac{\frac{7}{2}(p_2 V_2 - p_2 V_1)}{p_1 V_1 \ln \frac{V_2}{V_1} + \frac{5}{2} V_1 (p_1 - p_2)}$$

Questions and Problems

1. A copper cup of mass 0.300 kg contains 1.00 kg of water at 315 K. It is cooled so that the temperature decreases by 2.50 K each minute. What is the rate of removal of thermal energy in watts? The specific heat of water is 4,186 J/(kg · ℃) and the specific heat of copper is 387 J/(kg · ℃). _____.

    (A) 122   (B) 185   (C) 179   (D) 162

2. A simple pendulum having a length of 2 meters and a mass of 5 kg is pulled 30° from the vertical and released. If it is allowed to swing until it comes to rest, how much energy in J was converted into heat? _____.

    (A) 49   (B) 63   (C) 13   (D) 17

3. In an adiabatic free expansion _____.

    (A) no heat is transferred between a system and its surroundings

    (B) the pressure remains constant

    (C) the temperature remains constant

    (D) the process is reversible

4. The specific heat of liquid A is twice the specific heat of liquid B. The same amount of thermal energy is added to equal masses of the two fluids at the same initial temperature. Compare the change in temperature of A, with that of B. _____.

    (A) $\Delta T_A = \Delta T_B = 0$ (B) $\Delta T_A = \Delta T_B$

    (C) $\Delta T_A = 2\Delta T_B$ (D) $\Delta T_B = 2\Delta T_A$

5. In Fig. 8-7, 2 moles of an ideal gas start in initial state a, double in volume at constant pressure, change pressure at constant volume, decrease in volume to the original volume and then return to the original pressure. The work done on the gas in J in process is _____.

    (A) -20 (B) $+20$ (C) $+40$ (D) $+60$

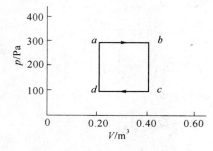

Fig. 8-7 Problem 5

Chapter 8 The First Law of Thermodynamics

6. In a cyclic process, which expression is correct? _____.
   (A) $\Delta U=0$ (B) $Q=-W$
   (C) $\Delta T=0$ (D) all of the above are correct

7. A 25 g lead bullet at 0℃ moves at 375 m/s and strikes a block of ice at 0℃. The quantity of ice _____ kg is melted if all of the kinetic energy of the bullet is converted to heat? The block of ice does not move(The latent heat of fusion of ice is 334.944 kJ/kg and the specific heat of lead is 0.128 kJ/(kg · ℃)).

8. 2.5 g of water at 100 K occupies a volume of 2.5 cm³ at atmospheric pressure. When the water is boiled, it occupies 4,178 cm³ as steam. The change in its internal energy is _____ J. The latent heat of vaporization is 2.26×10^6 J/kg. The pressure of one atm is 1.013×10^5 N/m².

9. Gas in a container increases its pressure from 101.325 kPa to 303.975 kPa while keeping its volume constant. The work done by the gas is _____ J, if the volume is 5 L($R=$ 8.31 J/(mol · K)).

10. 0.051 kg of metal at 250 ℃ is put into a calorimeter containing 0.41 kg of water at 25 ℃. The final temperature of the system is 30 ℃. The specific heat of the metal is _____ J/(kg · ℃)? Ignore any thermal transfers between the calorimeter and its contents. The specific heat of water is 4,186 J/(kg · ℃).

11. The specific heat of silver is 234 J/(kg · ℃). If a silver bullet of mass 4 g is shot into an insulating material with a speed of 300 m/s and comes to rest, the temperature increase of the bullet is _____ ℃.

12. Five moles of an ideal gas expand isothermally at 100 ℃ to five times its initial volume. The heat is _____ J that flows into the system($R=8.31$ J/(mol · K)).

13. As shown in Fig. 8-8, in the process of taking a gas from state a to state c along the curved path, 85 J of heat leaves the system and 55 J of work is done on the system. (1) Determine the change in internal energy, U_a-U_c. (2) When the gas is taken along the path cda, the work done by the gas is 38 J. How much heat Q is added to the gas in the process cda? (3) If $P_a=2.2P_d$, how much work is done by the gas in the process abc? (4) What is Q for path abc? (5) If , what is Q for the process bc? Here is a summary of what is given:
$Q_{\text{curve } a \to c}=-85$ J, $W_{\text{curve } a \to c}=-55$ J, $W_{cda}=38$ J, $U_a-U_b=15$ J, $p_a=2.2p_d$.

Fig. 8-8 Problem 13 and 14

14. Suppose a gas is taken clockwise around the rectangular cycle shown in Fig. 8-8,

starting at b, then to a, to d, to c, and returning to b. Using the values given in Problem 13, (1) describe each leg of the process, and then calculate (2) the net work done by the gas during the cycle, (3) the total internal energy change during the cycle, and (4) the net heat flow during the cycle. (5) What percentage of the intake heat was turned into usable work: i.e., how efficient is the "rectangular" cycle (given as a percentage)?

Chapter 9 The Second Law of Thermodynamics

Review of the Contents

1. Reversible and irreversible processes

There are two main types of thermodynamic processes: reversible process and irreversible process. The reversible process is the ideal process which never occurs; the irreversible process in Fig. 9-1 is the natural process which is commonly found in the nature.

In a reversible process, the system undergoing the process can be returned to its initial conditions along the same path on a p-V diagram, and every point along this path is an equilibrium state. A process that does not satisfy this requirement is irreversible.

2. The second law of thermodynamics

The first law of thermodynamics is very important, however, it makes no distinction between process that occurs spontaneously and those that do not. The second law of thermodynamics establishes which processes do and which do not occur. This law can be stated in various ways, including:

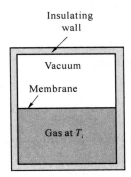

Fig. 9-1 The adiabatic free expansion of a gas is an irreversible process

(1) The Kelvin-Planck statement of the second law of thermodynamics

It's impossible to construct a heat engine that, operating in a cycle, produces no effect other than the input of energy by heat from a reservoir and the performance of an equal amount of work.

(2) The Clausius statement of the second law of thermodynamics

It's impossible to construct a cyclical machine whose sole effect is to transfer energy continuously by heat from one object to another object at a higher temperature without the input of energy by work.

All these versions are proved to be equal, and they are used to explain the phenomenon of irreversibility in nature.

3. Entropy

(1) Carnot's theorem

Carnot engine is of great importance from both practical and theoretical viewpoints. It operates in an ideal, reversible cycle, that is, Carnot cycle. This ideal engine establishes an upper limit on the efficiencies of all other engines. Carnot's theorem can be stated as follows:

All reversible engines operating between the same two constant temperatures T_H and T_L have the same efficiency. Any irreversible engine operating between the same two fixed temperatures will have an efficiency less than that of reversible engines.

The thermal efficiency of the engine is given by

$$e = \frac{W}{|Q_H|} = \frac{|Q_H| - |Q_L|}{|Q_H|} = 1 - \frac{|Q_L|}{|Q_H|} \tag{9-1}$$

The thermal efficiency of a Carnot engine is

$$e_C = 1 - \frac{T_L}{T_H} \tag{9-2}$$

This result indicates that all Carnot engines operating between the same two temperatures have the same efficiency.

(2) Clausius entropy

Entropy was originally formulated as a useful concept in thermodynamics. In 1865, Clausius first gave a mathematical version of the concept of entropy, and gave it its name.

The original formulation of entropy in thermodynamics involves the transfer of energy by heat during a reversible process. The definition of Clausius entropy in any infinitesimal process is expressed as follows:

$$dS = \frac{dQ}{T} \tag{9-3}$$

Entropy S is a state function of the system, and it depends only on the state property of the system.

In a reversible process, the system follows an arbitrary reversible process between the same initial and final states as the irreversible process,

$$\Delta S = S_b - S_a = \int_a^b dS = \int_a^b \frac{dQ}{T} \tag{9-4}$$

This equation means that the change in entropy during a process depends only on the end points and therefore is independent of the actual path followed. Consequently, the entropy change for an irreversible process can be obtained by calculating the entropy change for a reversible process that connects the same initial and final states.

If a system takes through an arbitrary reversible cycle in Fig. 9-2, for the entropy depends only on the properties of a given equilibrium state, then the change of the entropy in this process is

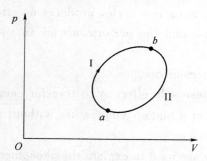

Fig. 9-2 A reversible cycle in p-V diagram

$$\Delta S = \oint_R dS = \oint_R \frac{dQ}{T} = 0$$

where the integration is over a closed path.

It is well known that the change in entropy for a system and its surroundings is always

Chapter 9 The Second Law of Thermodynamics

positive for an irreversible process. In general, the total entropy always increases in an irreversible process. This can be expressed as the following equation.

$$dS \geq \frac{dQ}{T} \tag{9-5}$$

Form this point of view, the second law of thermodynamics can be stated as follows:
The total entropy of an isolated system that undergoes a change cannot decrease.

4. Boltzmann entropy from a statistical view

The entropy of a system can be defined in terms of the possible distributions of its molecules. For identical molecules, each possible distribution of molecules is called a microstate of the system. All equivalent microstates are grouped into a configuration of the system. The number of microstates in a configuration is multiplicity W of the configuration.

For a system of N molecules that may be distributed between the two halves of a box, the multiplicity is given by

$$W = \frac{N!}{n_1! \, n_2!} \tag{9-6}$$

in which n_1 is the number of molecules in one half of the box and n_2 is the number in the other half. A basic assumption of statistical mechanics is that all the microstates are equally probable. Thus configurations with a large multiplicity occur most often. When N is very large, the molecules are nearly always in the configuration in which $n_1 = n_2$.

The multiplicity W of a configuration of a system and the entropy S of the system in that configuration are related by Boltzmann's entropy equation:

$$S = k \ln W \tag{9-7}$$

where $k = 1.38 \times 10^{-23}$ J/K is the Boltzmann constant.

When N is very large, we can approximate $\ln N$ with Stirling's approximation:

$$\ln N! \approx N \ln N - N \tag{9-8}$$

Typical Examples

Problem solving strategy

The change of entropy in different thermodynamics processes

Entropy is a state variable, ΔS depends only on the initial and final state of the system. For an irreversible process, we can figure out some reversible process that takes the system between the same two states.

Writing the first law of thermodynamics in differential form,

$$dQ = dU + dW \tag{9-9}$$

For an ideal gas, we have $pV = nRT$. Therefore, we can express heat energy as

$$dQ = nC_{V,m} dT + pdV \tag{9-10}$$

So, the entropy change for an ideal gas in the infinitesimal process can be obtained by

$$dS = \frac{pdV + dU}{T} = nC_{V,m} \frac{dT}{T} + nR \frac{dV}{V} \tag{9-11}$$

The entropy change ΔS is obtained by integration,

$$\Delta S = \int_{V_1}^{V_2} \frac{nR}{V} dV + \int_{T_1}^{T_2} \frac{nC_{V,m}}{T} dT$$

$$= nR \ln \frac{V_2}{V_1} + nC_{V,m} \ln \frac{T_2}{T_1} \qquad (9-12)$$

(1) A reversible isothermal process

$$\Delta S = nR \ln \frac{V_2}{V_1} \qquad (9-13)$$

(2) A reversible isochoric process

$$\Delta S = nC_{V,m} \ln \frac{T_2}{T_1} \qquad (9-14)$$

(3) A free expansion process

This is an irreversible process. We can design an isothermal process to connect the initial and final state. The total change of entropy can be expressed as follows:

$$\Delta S = nR \ln \frac{V_2}{V_1} \qquad (9-15)$$

Examples

1. A Carnot engine operates between the temperatures $T_H = 850$ K and $T_L = 300$ K. The engine performs 1,200 J of work each cycle, which takes 0.25 s.

(1) What is the efficiency of this engine?

(2) What is the average power of this engine?

(3) How much energy $|Q_H|$ is extracted as heat from the high temperature reservoir every cycle?

(4) How much energy $|Q_L|$ is delivered as heat to the low temperature reservoir every cycle?

(5) What is the entropy change of the working substance in this process?

Solution:

(1) The efficiency of a Carnot engine is

$$\varepsilon = 1 - \frac{T_L}{T_H} = 1 - \frac{300 \text{ K}}{850 \text{ K}} = 0.647 \approx 65\%$$

(2) The average power of an engine is defined as follows.

$$P = \frac{W}{t} = \frac{1,200 \text{ J}}{0.25 \text{ s}} = 4,800 \text{ W} = 4.8 \text{ kW}$$

(3) The efficiency is also connected with the work W and heat energy that is absorbed from the high temperature reservoir.

$$|Q_H| = \frac{W}{\varepsilon} = \frac{1,200 \text{ J}}{0.647} = 1,855 \text{ J}$$

(4) For a Carnot engine, the heat energy transfers to low temperature reservoir is

$$|Q_L| = |Q_H| - W = 1,855 \text{ J} - 1,200 \text{ J} = 655 \text{ J}$$

(5) For the high-temperature reservoir at T_H, the change in the entropy of the working substance is

Chapter 9 The Second Law of Thermodynamics

$$\Delta S_H = \frac{Q_H}{T_H} = \frac{1,855 \text{ J}}{850 \text{ K}} = 2.18 \text{ J/K}$$

Similarly, for the negative entropy transfer of energy to the low-temperature reservoir, we have

$$\Delta S_L = \frac{Q_L}{T_L} = \frac{-655 \text{ J}}{300 \text{ K}} = -2.18 \text{ J/K}$$

So the net entropy change of the working substance for one cycle is zero.

2. Suppose that there are 100 indistinguishable molecules in the box, as shown in Fig. 9-3. How many microstates are associated with the configuration $n_1 = 50$ and $n_2 = 50$? How many are associated with the configuration $n_1 = 100$ and $n_2 = 0$?

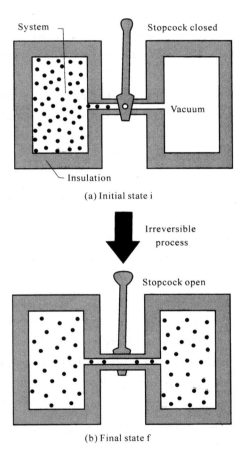

(a) Initial state i

Irreversible process

(b) Final state f

Fig. 9-3 Example 2

Solution:

The multiplicity W of a configuration of indistinguishable molecules in a closed box is the number of independent microstates with that configuration. For the (n_1, n_2) configuration (50, 50), that equation yields

$$W = \frac{N!}{n_1! \, n_2!} = \frac{100!}{50! \, 50!} = \frac{9.33 \times 10^{157}}{(3.04 \times 10^{64})(3.04 \times 10^{64})} = 1.01 \times 10^{29}$$

Similarly, for the configuration of (100, 0), we have

· 107 ·

$$W = \frac{N!}{n_1! \, n_2!} = \frac{100!}{100! \, 0!} = \frac{1}{0!} = \frac{1}{1} = 1$$

Questions and Problems

1. Thermal efficiency can be described mathematically as _____.

   (A) $\dfrac{Q_1 Q_2}{Q_3}$ (B) $\dfrac{W}{Q_H}$ (C) $\dfrac{Q_H + Q_L}{Q_H}$ (D) $1 - \dfrac{Q_H}{Q_L}$

2. What is the change in entropy in J/K of 2.51 mol of an ideal gas due to a free expansion to 3.62 times its original volume? _____.

   (A) 25.2   (B) 24.1   (C) 27.3   (D) 26.8

3. 1.23 kg of water at 273 K is mixed with 1.23 kg of water at 373 K. 323 K is the equilibrium temperature. What is the change in entropy in J/K of the system if $c = 4,186$ J/(kg·K)? _____.

   (A) 125   (B) 110   (C) −740   (D) 360

4. If 5 moles of an ideal gas at 0 ℃ are compressed isothermally from an initial volume of 100 cm^3 to a final volume of 20 cm^3, the change in entropy in J/K is _____.

   (A) −52   (B) −71   (C) −67   (D) −208

5. A company that produces pulsed gas heaters claims its efficiency is approximately 90%. If the engine operates between 250 ℃ and 25 ℃, what percentage is its thermodynamic efficiency? _____.

   (A) 83%   (B) 43%   (C) 90%   (D) 56%

6. A new electric power plant has an efficiency of 42%. For every 100 barrels of oil needed to run the turbine, how many are essentially lost as waste heat to the environment in barrels of oil? _____.

   (A) 21   (B) 42   (C) 58   (D) 10

7. An ice-making machine operates in a Carnot cycle. It takes water at 0 ℃ and rejects heat into the room at 27 ℃. If the machine produces 1 kg of ice, and the engine cycles 1,000 times during this process, the change in entropy in 4.186,8 J/K during one cycle is _____.

   (A) 0.25   (B) 0.29   (C) 0.27   (D) 0

8. The second law of thermodynamics states that _____.

   (A) it is impossible to construct a heat engine that, operating in a cycle, produces no other effect than absorption of energy from a reservoir and the performance of an equal amount of work

   (B) energy does not flow spontaneously from a cold object to a hot object

   (C) the entropy of the universe increases in all natural processes

   (D) all of the above are correct

9. 923 J of heat is transferred from a body at 383 K to a body at 283 K. What is the minimum change in entropy is _____ J/K?

10. 10 kg of water at 0 ℃ is mixed with 10 kg of water at 100 ℃. The change in entropy

of the system is _____ kJ/K.

11. Five moles of an ideal gas is allowed to undergo a free expansion. If the initial volume is 20 cm³ and the final volume is 100 cm³, the change in entropy is _____ J/K, if $R = 8.31$ J/(mol·K).

12. An engine is designed to obtain energy from the temperature gradient of the ocean. The thermodynamic efficiency of such an engine is _____, if the temperature of the surface of the water is 15 ℃ and the temperature well below the surface is 5 ℃.

13. A heat pump has a coefficient of performance of 4. If the heat pump absorbs 83.74 J of heat from the cold outdoors in each cycle, the heat expelled to the warm indoors is _____ J.

14. The solar constant, the radiant energy received from the sun per unit area per second, is 1.4 kW/m². The Earth's mean radius is 6.37×10^6 m. The efficiency of the atmosphere as a heat engine is 0.8%. The energy that the atmosphere converts into prevailing winds in a 24 h period of time is approximately _____.

15. The minimum change in entropy in the universe in J/K when a human body processes a 8.4 kJ daily diet at a body temperature of 37 ℃ is _____.

16. The Stirling cycle, shown in Fig. 9-4 is useful to describe external combustion engines as well as solar-power systems, assuming a monatomic gas as the working substance. The process *ab* and *cd* are isothermal whereas *bc* and *da* are at constant volume. How does it complete to the Carnot efficiency?

17. One mole of an ideal monatomic gas at STP(standard temperature and pressure) first undergoes an isothermal expansion so that the volume at *b* is 2.5 times the volume at *a* (Fig. 9-5). Next, heat is extracted at a constant volume so that the pressure drops. The gas is then compressed adiabatically back to the original state. (1) Calculating the pressure at *b* and *c*. (2) Determine the temperature at *c*. (3) Determine the work done, heat input or extracted, and the change in entropy for each process. (4) What is the efficiency of this cycle?

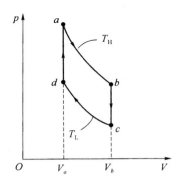

Fig. 9-4 Problem 16

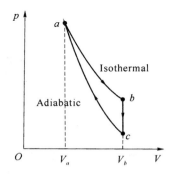

Fig. 9-5 Problem 17

Chapter 10 Oscillations

Review of the Contents

1. Simple harmonic motion

Periodic motion is a kind of motion of an object that repeats regularly. The object returns to a given position after a fixed time interval.

The block-spring system shown here is a classic linear simple harmonic oscillator where "linear" means that the force F is proportional to the displacement x rather than to some other power of x.

(1) Mathematical representation

In Fig. 10-1, (a), (b) and (c) show the block in different positions, the force exerted on the block is given by Hooke's law and is proportional to the position,

$$F = -kx \tag{10-1}$$

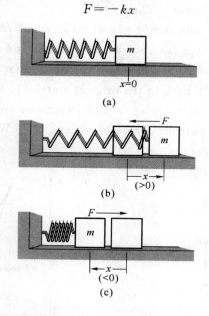

Fig. 10-1 A block attached to a spring moving on a frictionless table

We can write this equation in differential form,

$$\frac{d^2 x}{dt^2} + \omega^2 x = 0 \tag{10-2}$$

where $\omega = \sqrt{\dfrac{k}{m}}$.

The following cosine function is a solution to the differential equation.

· 110 ·

Chapter 10　Oscillations

$$x = A\cos(\omega t + \phi) \tag{10-3}$$

It is a mathematical representation of the position of the particle as a function of time, which is also called the motional equation of a simple harmonic oscillator.

(2) The characteristic quantities

Amplitude A is the maximum value of the position of the particle in either the positive or negative x direction. It is determined by the initial condition of the oscillation system.

$$A = \sqrt{x_0^2 + \frac{v_0^2}{\omega^2}} \tag{10-4}$$

Angular frequency ω is a measure of how rapidly the oscillations are occurring. It is defined by

$$\omega = \sqrt{\frac{k}{m}} \tag{10-5}$$

It can be written in the form of frequency or period.

$$\omega = 2\pi f = \frac{2\pi}{T} \tag{10-6}$$

The constant angle ϕ is called initial phase angle, and it can be obtained uniquely by the position and velocity of the particle at $t=0$.

$$\phi = \arctan\left(-\frac{v_0}{\omega x_0}\right) \tag{10-7}$$

The quantity $(\omega t + \phi)$ is called the phase of the motion. It will reflect the motion state of the particle at different time.

(3) Energy of simple harmonic oscillator

The total mechanical energy of a simple harmonic oscillator is a constant of the motion. The kinetic energy of the particle in Fig. 10-2 can be represented by

$$K = \frac{1}{2}mv^2 = \frac{1}{2}m\omega^2 A^2 \sin^2(\omega t + \phi) \tag{10-8}$$

The potential energy in Fig. 10-2 stored in the system for any elongation x is given by

$$U = \frac{1}{2}kx^2 = \frac{1}{2}kA^2\cos^2(\omega t + \phi) = \frac{1}{2}m\omega^2 A^2 \cos^2(\omega t + \phi) \tag{10-9}$$

The total mechanical energy is

$$E = K + U = \frac{1}{2}kA^2 \tag{10-10}$$

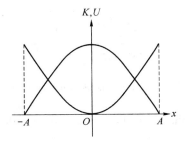

Fig. 10-2　Kinetic energy and potential energy versus position for a simple harmonic oscillator

2. Circle of reference or phasor

In our daily life, we can find that some common devices exhibit a relationship between oscillatory motion and circular motion, such as the pistons' motion in an automobile engine. It is also very convenient to introduce the circular motion to obtain the phase or initial phase of the simple harmonic oscillator.

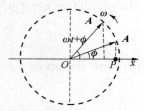

Fig. 10-3 Relationship between the uniform circular motion and the simple harmonic oscillator

In Fig. 10-3, a particle moves in a circle of radius A with constant angular speed ω. Its projection P on x axis behaves as a simple harmonic motion. At $t=0$, the initial phase angle ϕ is the angle between the vector $\mathbf{A}$ and the x axis. At some time $t>0$, the angle between $\mathbf{A}$ and the x axis is $\theta=\omega t+\phi$. We can conclude that the x coordinate of point P is

$$x = A\cos(\omega t + \phi) \qquad (10\text{-}11)$$

Simple harmonic motion along a straight line can be represented by the projection of uniform circular motion along a diameter of a reference circle.

3. Superposition of SHM

Generally, there are only a few examples of the ubiquity of harmonic motion. Most of oscillatory motion can be treated with the principle of superposition.

Consider the superposition of two simple harmonic motions in Fig. 10-4, which produce a displacement of the particle along the same line. Suppose that both have the same frequency, the resultant motion is also a simple harmonic motion with the same frequency.

$$x_1 = A_1\cos(\omega t + \phi_1) \qquad (10\text{-}12)$$
$$x_2 = A_2\cos(\omega t + \phi_2) \qquad (10\text{-}13)$$

The resulting displacement is given by:

$$x = x_1 + x_2 = A\cos(\omega t + \phi) \qquad (10\text{-}14)$$

where amplitude and initial phase can be determined by using the circle of reference.

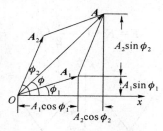

Fig. 10-4 The circle of reference is used to determine the amplitude and initial phase of the resultant simple harmonic motion

The amplitude and initial phase can be expressed as follows, respectively.

$$A = \sqrt{A_1^2 + A_2^2 + 2A_1 A_2 \cos(\phi_2 - \phi_1)} \qquad (10\text{-}15)$$

$$\phi = \arctan\frac{A_1\sin\phi_1 + A_2\sin\phi_2}{A_1\cos\phi_1 + A_2\cos\phi_2} \qquad (10\text{-}16)$$

Chapter 10   Oscillations

If the two simple harmonic motions have different frequencies, or different vibration direction, the resultant motion is very complex. However, if we have two vertical simple harmonic motions, it is possible to get Lissajous curve, as shown in Fig. 10-5.

$$x = A_1 \cos(3\omega t)$$

$$y = A_2 \cos\left(2\omega t + \frac{\pi}{4}\right)$$

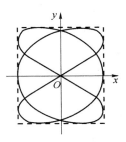

Fig. 10-5 A Lissajous curve is superposed by two vertical simple harmonic motions

Typical Examples

Problem solving strategy

1. The most of the problems concerned with the simple harmonic oscillator are really finding the motional equation (10-3) of them. Finding the equation of (10-3) is equivalent to finding the three characteristic quantities ω, A, ϕ. A and ϕ are determined by initial conditions as

$$A = \sqrt{x_0^2 + \frac{v_0^2}{\omega^2}}, \quad \phi = \arctan\left(-\frac{v_0}{\omega x_0}\right)$$

ω is related to the essential nature of the oscillator, and determined by the essential natures of the oscillator.

2. The phasor diagram described in section 2 of review of the contents is a very useful tool that makes the oscillation problems visualized, and is suggested to draw when you face a problem connected with an oscillation.

3. Since the total mechanical energy of an oscillator is conserved, we can find the differential equation by calculating the total mechanical energy of the oscillator. For example, for a block-spring system shown in Fig. 10-1, we can calculate its total mechanical enegy as

$$E = \frac{1}{2}mv^2 + \frac{1}{2}kx^2 = \text{constant}$$

We can obtain a relation as follows by taking derivative at both sides of the equation

$$mv\frac{dv}{dt} + kx\frac{dx}{dt} = 0$$

and

$$\frac{d^2x}{dt^2} + \frac{k}{m}x = 0$$

which is exactly the equation (10-2).

Examples

1. An object of mass 4 kg is attached to a spring of $k = 100$ N/m. The object is given an

initial velocity of $v_0 = -5$ m/s and an initial displacement of $x_0 = 1$ m. Find the motional equation of the object.

Solution:

The general motion equation for simple harmonic oscillator is $x = A\cos(\omega t + \phi)$, where $\omega = \sqrt{\dfrac{k}{m}} = \sqrt{\dfrac{100}{4}} = 5$ rad/s and $A = \sqrt{x_0^2 + \dfrac{v_0^2}{\omega^2}} = 1.4$ m respectively.

The initial phase/phase constant can be solved by using the initial conditions.

$$\tan\phi = -\dfrac{v_0}{\omega x_0} = 1$$

Besides, $v_0 = -\omega A\sin\phi$ is negative, so the initial phase is

$$\phi = \dfrac{\pi}{4}$$

The simple harmonic motion of the object is

$$x = 1.4\cos\left(5t + \dfrac{\pi}{4}\right)$$

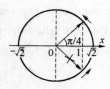

Fig. 10-6 Example 1

By the way, to calculate the initial phase, it is very useful to apply for the circle of reference(Fig. 10-6).

We can see that in the circle of reference, the initial phase has to choose $\dfrac{\pi}{4}$, in order to satisfy the initial direction of the object along the x-axis.

2. A particle undergoes SHM with $A = 4$ cm, $f = 0.5$ Hz. The displacement $x = -2$ cm when $t = 1$ s, and is moving along the positive x-axis. Write the simple harmonic motion equation.

Solution: From the description of the problems, we can see that $A = 4$ cm and $f = 0.5$ Hz. Therefore

$$\omega = 2\pi f = \pi \text{ rad/s}$$

The expression of simple harmonic motion is

$$x = 0.04\cos(\pi t + \phi)$$

Using the circle of reference, we can obtain the initial phase $\phi = \dfrac{\pi}{3}$ in Fig. 10-7.

The equation of SHM is $x = 0.04\cos\left(\pi t + \dfrac{\pi}{3}\right)$.

Fig. 10-7 Example 2

Questions and Problems

1. What is the maximum value of the velocity v when $x = A\cos(\omega t + \phi)$? _____.
   (A) ω (B) $\omega^2 A$ (C) $A\phi$ (D) ωA

2. The motion of a particle is described by $x = 10\sin(\pi t + \pi/3)$. At what time in s is the potential energy equal to the kinetic energy? _____.

Chapter 10 Oscillations

(A) 0.7   (B) 0.8   (C) 0.9   (D) 0.6

3. A uniform rod (length $L=1.0$ m, mass$=2.0$ kg) is suspended from a pivot a distance $d=0.25$ m above its center of mass. The angular frequency in rad/s for small oscillations is approximately _____.

(A) 1.0   (B) 2.2   (C) 1.5   (D) 4.1

4. The mass attached to a spring is replaced by a mass four times as large. By what factor is the period of the spring changed? _____.

(A) The period doesn't change   (B) 16
(C) 4   (D) 2

5. Fig. 10-8 shows a pendulum at a point in its swing at which it is moving with velocity v. Which arrow correctly shows the direction of its acceleration? _____.

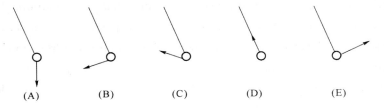

Fig. 10-8   Problem 5

6. When the position of an oscillating particle is $x = A\cos \omega t$, its acceleration a is _____.

7. A body of mass 5 kg stretches a spring 10 cm when the mass is attached. It is then displaced downward an additional 5 cm and released. Its position as a function of time in cm is approximately _____. (SI)

8. A horizontal plank ($m=2.0$ kg, $L=1.0$ m) is pivoted at one end. A spring($k=1,000$ N/m) is attached at the other end, as shown in Fig. 10-9. Find the angular frequency in rad/s for small oscillations is _____.

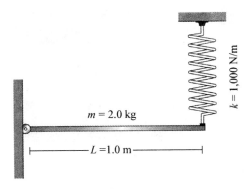

Fig. 10-9   Problem 8

9. A mass is connected to two identical springs as shown in Fig. 10-10. Each spring has spring constant k. When the mass is pulled to one side and released, its frequency of

· 115 ·

vibration is _____.

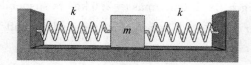

Fig. 10-10   Problem 9

10. A block on the end of a spring is at rest at $x=x_e$. It is pulled out to position $x=x_e+A$ and released. The distance it travels in one full cycle of its motion is _____.

11. Two 500 g masses sitting on a frictionless surface are attached to the ends of a spring with spring constant 4,000 N/m. The spring is compressed 6.00 cm and then released suddenly. The period of vibration of the system is _____ ms.

12. A particle attached to a spring executes simple harmonic motion. When it passes through the equilibrium position it has a speed v. The particle is stopped, and then the oscillation is restarted so that it now passes through the equilibrium position with a speed of $2v$. After this change, the frequency of oscillation will change by a factor of _____, the amplitude of the oscillation will change by a factor of _____, the magnitude of maximum acceleration of particle will change by a factor of _____.

13. A mass m is connected to two springs, with spring constants k_1 and k_2, in two different ways as shown in Fig. 10-11(a) and (b). Show that the period for the configuration shown in part(a) is given by

$$T=2\pi\sqrt{m\left(\frac{1}{k_1}+\frac{1}{k_2}\right)}$$

And for that in part (b) is given by

$$T=2\pi\sqrt{\frac{m}{k_1+k_2}}$$

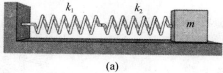

(a)

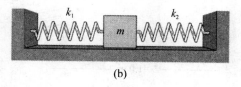

(b)

Fig. 10-11   Problem 13

14. A spring with spring constant 250 N/m vibrates with amplitude of 12.0 cm when 0.380 kg hangs from it. (1) What is the equation describing this motion as a function of time? Assume the mass passes through the equilibrium point, toward positive x (upward), at $t=0.110$ s. (2) At what times will the spring have its maximum and minimum lengths?

(3) What is the displacement at $t=0$? (4) What is the force exerted by the spring at $t=0$? (5) What is the maximum speed and when is it first reached after $t=0$?

15. Imagine that a 10 cm diameter circular hole were drilled all the way through the center of the Earth (Fig. 10-12). At one end of the hole, you drop an apple into the hole. Show that, if you assume that the Earth has a constant density, the subsequent motion of the apple is simple harmonic. How long will the apple take to return? Assume that we can ignore all frictional effects.

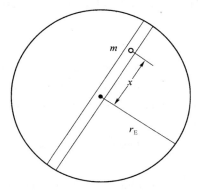

Fig. 10-12 Problem 15

Chapter 11 Wave Motion

Review of the Contents

1. Concepts of mechanical wave

A wave is a disturbance in a medium that travels outward from its source as shown in Fig. 11-1. It travels from one place to another by means of a medium, but the medium itself is not transported. All material media—solids, liquids, and gases—can carry energy and information by means of waves.

(1) Transverse wave: the particle displacement is perpendicular to the direction of wave propagation.

(2) Longitudinal wave: in a longitudinal wave the particle displacement is parallel to the direction of wave propagation.

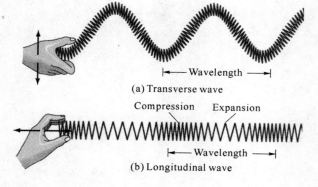

Fig. 11-1 The wave in spring system

(3) Wavelength (Fig. 11-2) is defined by the minimum distance between any two identical points on adjacent waves. The period is the time interval required for two identical points of adjacent waves to pass by a point. It is the same as the period of the simple harmonic oscillation. The speed of a wave is a property of the medium—changing the speed actually requires a change in the medium itself. If the medium does not change as a wave travels, the wave speed is constant. The wave speed, wavelength, and period are related by the expression

$$v = \frac{\lambda}{T}$$

(4) Plane wave (Fig. 11-3 (a)): the disturbance travels in single direction; plane represents wave front, which is composed of elements with same phase.

Chapter 11 Wave Motion

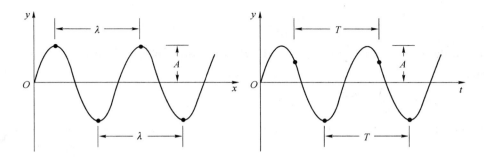

Fig. 11-2 The wavelength, period and amplitude are marked in the figures

(5) Spherical wave (Fig. 11-3(b)): the disturbance travels in radial direction; plane represents wave front, and the solid line with arrow means wave lines.

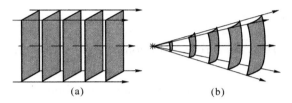

Fig. 11-3 A schematic representation of plane wave and spherical wave

2. Plane harmonic wave function

The wave function represents the displacement $y(x,t)$ of the wave at any chosen point x at any time t. The function $y(x,t)$ describes the actual shape of the wave.

The wave function of the plane harmonic wave in Fig. 11-4 can be represented by the following expressions,

$$y(x,t) = A\cos\left(\omega t \pm \frac{2\pi}{\lambda}x + \phi_0\right)$$

$$y(x,t) = A\cos\left[2\pi\left(\frac{t}{T} \pm \frac{x}{\lambda}\right) + \phi_0\right]$$

In these expressions, "+" sign means that the wave is moving along left side of x-axis. "−" sign means that the wave is moving along right side of the x-axis. $k = \frac{2\pi}{\lambda}$ is wave number.

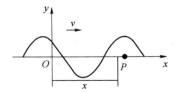

Fig. 11-4 A plane harmonic wave is traveling in the right direction along x-axis

3. Energy transfer in wave motion

Waves transport energy when they travel through a medium. In the process, the energy of particle in the elastic medium is associated with kinetic energy and potential energy.

The energy of unit volume (energy density) of the particle is given by

· 119 ·

$$w = \frac{dE}{dV} = \frac{dK + dU}{dV} = \rho\omega^2 A^2 \sin^2(\omega t - kx)$$

We can define that energy current is the flow of total energy through a cross section in unit time.

$$\text{energy current} = \frac{wV}{\Delta t} = \frac{wSv\Delta t}{\Delta t} = wvS$$

The energy current density is given in Fig. 11-5 as follows.

$$P = \frac{\text{energy current}}{S} = wv$$

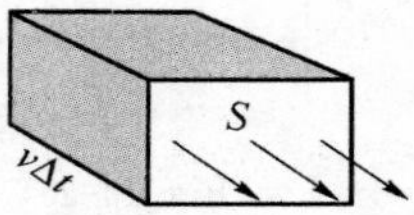

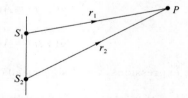

Fig. 11-5 A schematic show of energy current and energy density

Furthermore, we can get that the average of energy current density is

$$I = \overline{P} = \overline{w}v = \frac{1}{2}\rho\omega^2 A^2 v$$

Sometimes, we use $I = A^2$ as a representation of wave intensity.

4. Interference

In the case of mechanical waves, if two or more traveling waves are moving through a medium, the resultant wave function is the algebraic sum of the values of the wave functions of the individual waves. This is called superposition principle. Two traveling waves can pass through each other without being destroyed or even altered.

Interference in Fig. 11-6 is the addition of two or more waves that result in a new wave pattern. Interference usually refers to the interaction of waves that are correlated or coherent with each other. The source must be coherent, that is they must maintain a constant phase with respect to each other.

(a) Interference pattern produced with two identical circular waves

(b) The resultant oscillation at point P is superposition of oscillations stimulated by two wave sources at S_1 and S_2

Fig. 11-6 Interference by two wave sources

The oscillatory function at point P can be obtained by the principle of superposition.

$$y_1 = A_1 \cos\left(\omega t - \frac{2\pi}{\lambda}r_1 + \phi_1\right)$$

$$y_2 = A_2 \cos\left(\omega t - \frac{2\pi}{\lambda}r_2 + \phi_2\right)$$

$$y_P = y_1 + y_2 = A\cos(\omega t + \phi)$$

where A is the resultant amplitude and ϕ is the initial phase. Both A and ϕ can be determined by the following expressions,

Chapter 11 Wave Motion

$$A^2 = A_1^2 + A_2^2 + 2A_1 A_2 \cos \Delta\varphi$$

$$\Delta\varphi = (\phi_2 - \phi_1) - \frac{2\pi}{\lambda}(r_2 - r_1)$$

The intensity of wave is given by

$$I = I_1 + I_2 + 2\sqrt{I_1 I_2} \cos \Delta\varphi$$

If a crest of a wave meets a crest of another wave at the same point then the crests interfere constructively and the resultant wave amplitude is greater. If a crest of a wave meets a trough of another wave then they interfere destructively, and the overall amplitude is decreased.

For constructive interference, when $\Delta\varphi = \pm 2k\pi, k = 0, 1, 2, \cdots$, we can get

$$A = A_1 + A_2 \text{ and } I = I_1 + I_2 + 2\sqrt{I_1 I_2}$$

For destructive interference, when $\Delta\varphi = \pm(2k+1)\pi, k = 0, 1, 2, \cdots$, we can get

$$A = |A_1 - A_2| = A_{\min} \text{ and } I = I_1 + I_2 - 2\sqrt{I_1 I_2}$$

5. Standing wave

A standing wave in Fig. 11-7 can occur in a stationary medium as a result of interference between two identical waves traveling in opposite directions.

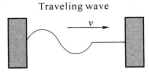

(a) A traveling wave in a tight string hits a boundary, and is reflected off it

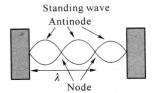

(b) Two identical waves would be traveling in the opposite direction from the initial wave, and would result in a standing wave

Fig. 11-7 Formation of a standing wave

(1) Standing wave function

Let the harmonic waves be represented by the equations below,

$$y_1 = A\cos\left(\omega t - \frac{2\pi}{\lambda}x\right)$$

$$y_2 = A\cos\left(\omega t + \frac{2\pi}{\lambda}x\right)$$

So the resultant wave equation will be the sum of y_1 and y_2,

$$y = y_1 + y_2 = 2A\cos\left(2\pi\frac{x}{\lambda}\right)\cos \omega t = A_{\text{stand}} \cos \omega t$$

This equation represents the wave function of a standing wave. It is in a kind of oscillation pattern.

(2) The feature of waveform

The behavior of the waves at the points of minimum and maximum vibrations (nodes and antinodes) contributes to the constructive interference which forms the resonant standing waves.

Node can be obtained by

$$\frac{2\pi}{\lambda}x = \pm(2k+1)\frac{\pi}{2}, x = \pm\left(k+\frac{1}{2}\right)\frac{\lambda}{2}, k=0,1,2,\cdots$$

Antinodes can be determined by

$$\frac{2\pi}{\lambda}x = \pm k\pi, x = \pm k\frac{\lambda}{2}, k=0,1,2,\cdots$$

The distance between adjacent antinodes/nodes is equal to $\frac{\lambda}{2}$.

The distance between a node and an adjacent antinode is $\frac{\lambda}{4}$.

(3) Normal modes

For string fixed at both ends in Fig. 11-8, the boundary condition results in the string having a number of natural patterns of oscillation. This is called normal modes. The wavelength of various normal modes is

$$\lambda_n = \frac{2L}{n}$$

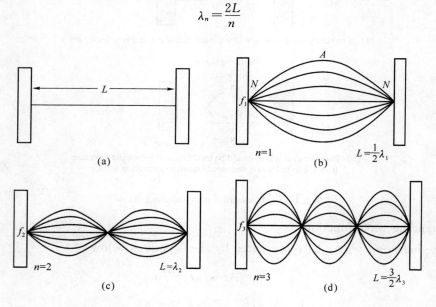

Fig. 11-8 A string of length L is fixed at both ends

The natural frequency associated with these wavelength are determined by

$$f_n = \frac{v}{\lambda_n} = \frac{n}{2L}v = \frac{n}{2L}\sqrt{\frac{T}{\mu}}, n=1,2,3,\cdots$$

Chapter 11 Wave Motion

Typical Examples

Problem solving strategy

Wave function and its solution.

Consider a transverse traveling wave in Fig. 11-9, the transverse position of elements can be described by
$$y(x,t)=f(x\mp vt)$$
where negative sign means that the wave travels to the right, and vice versa.

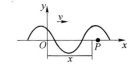

Fig. 11-9 A sketch of plane harmonic waveform

If we consider a plane harmonic wave, the vibration equation at point $x=0$ is given by
$$y(0,t)=A\cos(\omega t+\phi)$$

Then we can get the wave function from the view of phase or time relations.

The time delay between the origin and the point P is $\Delta t=\dfrac{x}{v}$; while the phase delay is given by $\Delta\phi=\omega\Delta t=\dfrac{2\pi}{T}\dfrac{x}{v}=\dfrac{2\pi}{\lambda}x$. Thus the wave function can be obtained in these two ways,

$$y_P(t)=A\cos\left[\omega\left(t-\dfrac{x}{v}\right)+\phi\right]$$

$$y_P(t)=A\cos\left[\omega t-\dfrac{2\pi}{\lambda}x+\phi\right]$$

It can be verified that these two equations are identical.

Examples

1. A plane harmonic wave in Fig. 11-10 travels in $+x$-direction with speed v and wavelength λ. The oscillatory function of particle at $x_0=\lambda/4$ is $y(x_0,t)=A\cos\omega t$. Find the wave function of this plane harmonic wave.

Solution: Consider a point P on the x-axis, its phase is retarded with respect x_0. The phase at point x is retarded with respect to x_0.

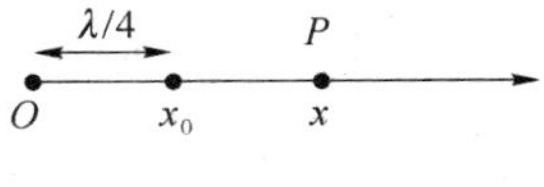

Fig. 11-10 Example 1

So the plane harmonic wave can be obtained using the standard wave equation,

$$y(x,t)=A\cos\left[\omega t-\dfrac{2\pi}{\lambda}(x-x_0)\right]$$
$$=A\cos\left[\dfrac{2\pi}{\lambda}vt-\dfrac{2\pi}{\lambda}\left(x-\dfrac{\lambda}{4}\right)\right]$$
$$=A\cos\left[\dfrac{2\pi}{\lambda}(vt-x)+\dfrac{\pi}{2}\right]$$

2. A harmonic wave travels in x-direction. The waveform at time $t=1$ s is shown in Fig. 11-11.

(1) Draw the direction of motion of particle marked with o, a, b, c.

(2) Write the wave function.

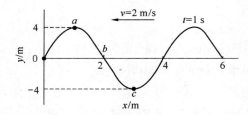

Fig. 11-11 Example 2

Solution: (1) The vibration direction of these particles are marked as follows.

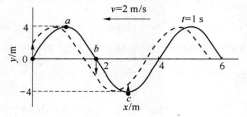

Fig. 11-12 Solution(1)

(2) In Fig. 11-12, we can obtain the following parameters. $A=4$ m, $\lambda=4$ m, and $\omega = 2\pi/T = \pi$ rad/s. The initial phase at point O is determined by the vibration state. Using the circle of reference, the phase angle of point O at $t=1$ s is $-\dfrac{\pi}{2}$, thus the initial phase at origin O is $\pi/2$ in Fig. 11-13 for one period is 2 s.

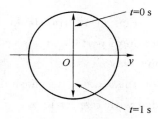

Fig. 11-13 Solution(2)

The wave function is $y(x,t) = 4\cos\left(\pi t + \dfrac{\pi}{2}x + \dfrac{\pi}{2}\right)$ (SI).

Questions and Problems

1. When a traveling wave reaches a boundary with a medium of greater density, the phase change in the reflected wave is _____.

    (A) 0   (B) $\pi/4$ (C) $\pi/2$ (D) π

Chapter 11 Wave Motion

2. Bats can detect small objects (such as insects) that are approximately the size of a wavelength. If bats emit a chirp at a frequency of 60 kHz and the speed of sound waves in air is 330 m/s, what is the smallest size insect they can detect in mm? _____.

(A) 1.6   (B) 3.5   (C) 5.5
(D) 7.5   (E) 9.8

3. The Superposition principle describes the observation that the wave displacement at a point is obtained by _____.

(A) adding the magnitudes of the two separate wave functions at that point
(B) subtracting the magnitudes of the two separate wave functions at that point
(C) multiplying the magnitudes of the two separate wave functions at that point
(D) adding the two separate wave functions at that point algebraically
(E) multiplying the two separate wave functions at that point algebraically

4. The wave function $y(x,t)$ represents _____.

(A) the position y at which the displacement x occurs at time t
(B) the amplitude of the waves for all points x at all times t
(C) the displacement y of any point P located at position x at time t
(D) the displacement y of the x coordinate axis at time t
(E) the wave speed x/t at position y

5. Constructive interference occurs at a point in space when the path difference for waves from coherent sources to that point is an integer multiple of _____.

(A) $\lambda/2$ (B) $\lambda/3$ (C) $\lambda/4$
(D) $\lambda/5$ (E) λ

6. Two harmonic waves are described by: $y_1 = 6 \sin [\pi(2x+3t)]$ cm and $y_2 = 6 \sin [\pi(2x-3t)]$ cm, with x in cm and t in s. Determine the smallest value of x in cm corresponding to a node. _____.

(A) 3   (B) 0.25   (C) 0
(D) 6   (E) 1.5

7. The wave functions of two waves on a linear medium are $y_1 = 12 \sin [\pi(3.0 x - 5.0 t)]$ cm and $y_2 = 12 \sin [\pi(3.0 x - 5.0 t) - 4.0 \text{ rad}]$ cm. What is the displacement y in cm at the point $x = 1.0$ m at time $t = 1.0$ s? _____.

(A) -3.46 (B) 0   (C) 2.80
(D) 9.08   (E) 12.0

8. The wave functions of two sound waves are $y_1 = 6.0 \sin [\pi(2.0 x + 3.0 t)]$ cm and $y_2 = 6.0 \sin [\pi(2.0 x - 3.0 t)]$ cm. What is the smallest positive value, in m, of the x-coordinate at which an antinode of a standing wave they create is located? _____.

(A) 0   (B) 0.25   (C) 0.50
(D) $\pi/2$ (E) $\pi/4$

9. For the wave described by $y = 0.15 \sin [(\pi/16)(2x - 64t)]$ (SI units), the maximum positive displacement in m occurs at $t=0$, when x is equal to _____.

10. The variation in the pressure of helium gas, measured from its equilibrium value, is given by $\Delta P = 2.9 \times 10^{-5} \cos (6.2x - 3,000 t)$ where x and t have units m and s. The

wavelength in m of the wave is _____ m.

11. A flute player holding a tone with a frequency of 520 Hz approaches a wall at 2 m/s on a day when the speed of sound in air is 340 m/s. The frequency in Hz he hears coming back to him from the wall is _____ Hz.

12. Two instruments produce a beat frequency of 5 Hz. If one has a frequency of 264 Hz, what could be the frequency of the other instrument _____ Hz?

13. Suppose two linear waves of equal amplitude and frequency have a phase difference ϕ as they travel in the same medium. They can be represented by

$$y_1 = A\sin(kx - \omega t)$$
$$y_2 = A\sin(kx - \omega t + \phi)$$

(1) Use the trigonometric identity $\sin\theta_1 + \sin\theta_2 = 2\sin\frac{1}{2}(\theta_1+\theta_2)\cos\frac{1}{2}(\theta_1-\theta_2)$ to show that the resultant wave is given by

$$y = \left(2A\cos\frac{\phi}{2}\right)\sin\left(kx - \omega t + \frac{\phi}{2}\right)$$

(2) What is the amplitude of this resultant wave? Is the wave purely sinusoidal, or not?

(3) Show that constructive interference occurs if $\phi = 0, 2\pi, 4\pi$ and so on, and destructive interference occurs if $\phi = \pi, 3\pi, 5\pi$, etc.

14. A standing wave on a 1.80 m long horizontal string displays three loops when the string vibrates at 120 Hz. The maximum swing of the string (top to bottom) at the center of each loop is 12.0 cm. (1) What is the function describing the standing wave? (2) What is the function describing the two equal-amplitude waves traveling in opposite directions that make up the standing wave?

15. In Fig. 11-14, a string, tied to a sinusoidal vibrator at P and running over a support at Q, is stretched by a block of mass m. The separation L between P and Q is 1.2 m, the linear density of the string is 1.6 g/m, and the frequency f of the vibrator is fixed at 120 Hz. The amplitude of the motion at P is small enough for that point to be considered a node. A node also exists at Q.

(1) What mass m allows the vibrator to set up the fourth harmonic on the string?

(2) What standing wave mode is set up if $m = 1.00$ kg?

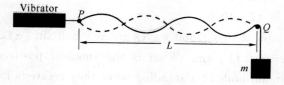

Fig. 11-14 Problem 15

Chapter 12 Interference of Light

Review of the Contents

1. Interference by two beam of light

As shown in Fig. 12-1, when two lightwaves emitted from the two sources S_1 and S_2 pass through a same region of space at the same time, they interfere. The resultant intensity at the point P in this region is:

$$I = I_1 + I_2 + 2\sqrt{I_1 I_2}\cos\Delta\varphi \tag{12-1}$$

where $\Delta\varphi$ is the phase difference at P between the two lightwaves.

$$\Delta\varphi = \frac{2\pi}{\lambda}(r_2 - r_1) - (\phi_2 - \phi_1) = \frac{2\pi}{\lambda}(r_2 - r_1), \text{suppose } \phi_2 = \phi_1 \tag{12-2}$$

Fig. 12-1 Interference by two light sources S_1 and S_2

where ϕ_1 and ϕ_2 are the phase angles (initial phases) of two sources S_1 and S_2.

At some point P

where $\Delta\varphi = \frac{2\pi}{\lambda}(r_2 - r_1) = \pm 2m\pi \quad (m = 0, 1, 2, \cdots)$ $\tag{12-3}$

We get the maximum intensity $I = I_1 + I_2 + 2\sqrt{I_1 I_2} \xrightarrow{I_1 = I_2} I_{max} = 4I_1$, whose cases are called constructive interference.

At some point P

where $\Delta\varphi = \frac{2\pi}{\lambda}(r_2 - r_1) = \pm(2m+1)\pi \quad (m = 0, 1, 2, \cdots)$ $\tag{12-4}$

We get the minimum intensity $I = I_1 + I_2 - 2\sqrt{I_1 I_2} \xrightarrow{I_1 = I_2} I_{min} = 0$, whose cases are called destructive interference.

2. Optical path length

As shown in Fig. 12-2, when a light wave travel across a series of medium composed of m layers, for medium i the index of refraction is n_i, the optical path length traversed by the wave is

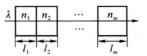

Fig. 12-2 A light beam getting through a multiple layer

$$L = \sum_{i=1}^{m} n_i l_i \tag{12-5}$$

The relationship between the optical path length and phase retardation:

$$\text{phase difference} = \Delta\varphi = \frac{2\pi}{\lambda}\delta = \frac{2\pi}{\lambda} \times \text{difference of optical path length} \tag{12-6}$$

3. Young's two-slit interference experiment

In Young's double-slit experiment (Fig. 12-3), two slits S_1 and S_2 separated by a distance d are illuminated by a monochromatic light source. An interference pattern consisting of bright and dark fringes is observed on a viewing screen.

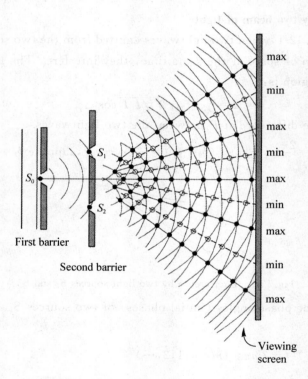

Fig. 12-3 Young's two-slit interference experiment

The condition for bright fringes (constructive interference) is

$$\delta = d\sin\theta = m\lambda \quad (m = 0, \pm 1, \pm 2, \cdots) \tag{12-7}$$

The condition for dark fringes (destructive interference) is

$$\delta = d\sin\theta = (2m+1)\frac{\lambda}{2} \quad (m = \pm 1, \pm 2, \cdots) \tag{12-8}$$

The intensity of interference pattern at a point in the viewing screen is

$$I = 4I_0 \cos^2\left(\frac{\pi d \sin\theta}{\lambda}\right) \tag{12-9}$$

where I_0 is the intensity when each slit is opened.

4. Coherence of light

The two slits in Young's experiment act as if they were two sources of radiation. They are called coherent sources because the waves leaving from the two slits bear the same phase

Chapter 12 Interference of Light

relationship to each other all the time. This phase relationship happens because the waves coming through the two slits come from a single narrow slit before the two slits. An interference pattern is clearly observed only when the sources are coherent (solid line in Fig. 12-4).

If two tiny lightbulbs replaced the two slits, an interference pattern would not be seen. The light emitted by one lightbulb would have a random phase with respect to the second bulb, and the screen would be more or less uniformly illuminated (dash line in Fig. 12-4). Two such sources, whose output waves have phases that bear no fixed relationship to each other over time, are called incoherent sources.

Actually when a human eye watches the pattern in the viewing screen, it makes a time average of the things it sees because the human eye has a response time τ to the pattern it sees. So the Eq. (12-1) will be modified by a time average over time τ which denoted by the symbol $<>$

$$I = I_1 + I_2 + 2\sqrt{I_1 I_2} <\cos \Delta\varphi> \qquad (12\text{-}10)$$

If two sources are incoherent, $<\cos \Delta\varphi> = 0$, $I = I_1 + I_2$, which means that the viewing screen is uniformly illuminated.

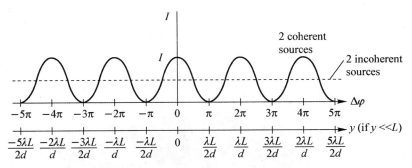

Fig. 12-4 Intensity I as a function of phase difference $\Delta\varphi$ and position on screen y (assuming $y \ll L$)

5. Thin-film interference

When light is incident on a thin transparent film, the lightwaves reflected from the front and back surfaces interfere. For near-normal incidence the condition for bright fringes (constructive interference) is

$$\delta = 2nt + \left(\frac{\lambda}{2}\right) = m\lambda \quad (m = 1, 2, 3, \cdots) \qquad (12\text{-}11)$$

The condition for dark fringes (destructive interference) is

$$\delta = 2nt + \left(\frac{\lambda}{2}\right) = (2m+1)\frac{\lambda}{2} \quad (m = 1, 2, 3, \cdots) \qquad (12\text{-}12)$$

where n is the index of refraction of the film, t is its thickness.

If the light incident at an interface between media with different indexes of refraction is in the medium with smaller index of refraction, the reflection causes a phase change of π rad, or half a wavelength, in reflected wave. Otherwise, there is no phase change due to the reflection. Refraction at an interface does not cause a phase shift. In the Eq. (12-11) and (12-12), $\lambda/2$ in the brackets is added or omitted according to the cases whether there are phase shifts in reflected waves at front and back surfaces of the film.

Fig. 12-5 Newton's ring

6. Newton's ring

A lens with a radius R is placed in contact with a flat glass surface, a series of concentric rings is seen when illuminated from above by a monochromatic light with wavelength λ. This phenomenon is called Newton's rings, as shown in Fig. 12-5.

The radii of bright rings are

$$r_m = \sqrt{\left(m - \frac{1}{2}\right)\lambda R} \quad (m=1,2,3,\cdots) \tag{12-13}$$

The radii of dark rings are

$$r_m = \sqrt{m\lambda R} \quad (m=0,1,2,\cdots) \tag{12-14}$$

7. Michelson interferometer

In Michelson's interferometer, a light wave is split into two beams (at the beamsplitter (Fig. 12-6)) that, after traversing paths of different lengths, recombined so they interfere and form a fringes pattern (at the viewing screen). Varying the path length of one of the beams allows distance d to be accurately expressed in terms of wavelength of light, by counting the numbers N of fringes through which the pattern shifts because of change.

$$d = N\frac{\lambda}{2} \tag{12-15}$$

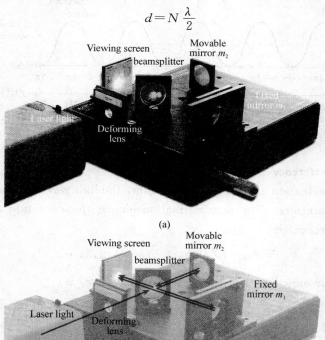

(a)

(b)

Fig. 12-6 Michelson interferometer

Chapter 12 Interference of Light

Typical Examples

Problem solving strategy

1. This chapter discusses the principle of interference of light, and some of the important instruments like the double-slit interference, the thin film, Michelson's interferometer, which make the phenomena of interference visible. In order to know which location the constructive interference (bright fringe) appears or which location the destructive interference (dark fringe) appears, we at first have to calculate the difference of optical path lengths δ between the two beams coming from the two coherent light sources. And the corresponding phase difference $\Delta\varphi = (2\pi/\lambda)\delta$.

For constructive interference: $\Delta\varphi = 2m\pi, m = 0, 1, 2, \cdots$,
$$\delta = m\lambda, m = 0, 1, 2, \cdots$$
For destructive interference: $\Delta\varphi = (2m+1)\pi, m = 0, 1, 2, \cdots$,
$$\delta = (2m+1)\frac{\lambda}{2}, m = 0, 1, 2, \cdots$$

2. If the light waves meet the interface of another medium to travel, we must consider whether we add $\lambda/2$ into the difference of optical path lengths according to the indexes of refraction in the two sides of the interface.

3. If a sheet of transparent material with thickness of t and index of refraction of n is inserted in one way of the beam, we have to take an additional optical path difference $(n-1)t$ into account just like the Example 1 as follows. For Michelson's interferometer, an inserted sheet will induce an additional optical path difference $2(n-1)t$ instead, for the reason that the light beams travel two times between the beam splitter and one of the mirrors (see Example 4).

Examples

1. Consider the double-slit arrangement shown in Fig. 12-7, where the slit separation is d and the slit to screen distance is L. A sheet of transparent plastic having an index of refraction n and thickness t is placed over the upper slit. As a result, the central maximum of the interference pattern moves upward a distance y'. Find y'.

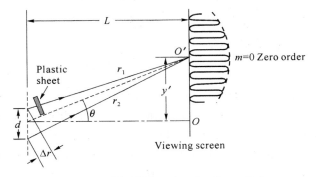

Fig. 12-7 The illustration for Example 1

Solution:

If this plastic sheet is not inserted, the central maximum (zero order bright fringe) would locate at O point, the center of the viewing screen. Now with the plastic sheet inserted, the zero order bright fringe will move upward a distance y', at the point O'. The difference of optical path length between two paths is

$$r_2 - [r_1 - t + nt] = r_2 - r_1 - (n-1)t = 0 \tag{12-16}$$

For the reason that θ is so small that we can make an approximation that

$$r_2 - r_1 \approx d\sin\theta \approx d\tan\theta = d\frac{y'}{L} \tag{12-17}$$

From Eq. (12-16) and (12-17) we get

$$d\frac{y'}{L} = (n-1)t, \text{ so } y' = \frac{(n-1)tL}{d} \tag{12-18}$$

2. An oil drop ($n = 1.20$) floats on a water ($n = 1.33$) surface and is observed from above by reflected light (see Fig. 12-8). (1) Will the outer (thinnest) regions of the drop correspond to a bright or a dark region? (2) How thick is the oil film where one observes the third blue region from the outside of the drop? (3) Why do the colors gradually disappear as the oil thickness becomes larger?

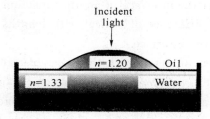

Fig. 12-8   The illustration for Example 2

Solution:

(1) There is no additional $\lambda/2$ term in the difference of optical path length $\delta = 2n_{oil}t$ for the reason of the indices on the sides of upper surface and bottom surface of the oil drop on the water surface. For the outer border of the drop, the thickness $t = 0$. Therefore $\delta = 0 \times \lambda$ which corresponds to the bright fringe.

(2) Let the average wavelength of the blue light be 450 nm. The thickness where one observes the third blue region from the outside of the drop satisfies the following equation

$$\delta = 2n_{oil}t = (3+1)\lambda_{blue} \tag{12-19}$$

$$t = \frac{4\lambda_{blue}}{2n_{oil}} = \frac{4 \times 0.450 \ \mu m}{2 \times 1.20} = 0.750 \ \mu m \tag{12-20}$$

(3) The distinct bright and dark fringes formed require that the two light beams coming from the reflection from upper and bottom surface of the oil drop are coherent. In this case, the length of the wavetrain of the light beams is larger than the optical path length difference. With the thickness of the oil film goes larger, the length of the wavetrain becomes smaller than the optical path length difference $\delta = 2n_{oil}t$, so that the two light beams are no longer coherent with each other. Therefore there is no interference pattern, or the

colors appeared on the surface disappeared.

3. The Fig. 12-9 illustrates a setup used for testing lenses. Show that:
$$d = \frac{x^2(R_2 - R_1)}{2R_1 R_2}$$

Prove that the radius of m-th dark fringe is then
$$x_m = \sqrt{\frac{R_1 R_2 m\lambda}{(R_2 - R_1)}}$$

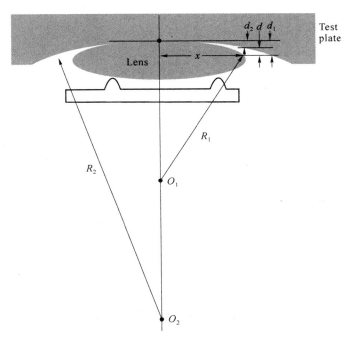

Fig. 12-9   The illustration for Example 3

Solution:

Let x be the radius of dark fringes, we have
$$x = \sqrt{2R_1 d_1} = \sqrt{2R_2 d_2} \tag{12-21}$$

So
$$d = d_1 - d_2 = \frac{x^2}{2}\left(\frac{1}{R_1} - \frac{1}{R_2}\right) = \frac{x^2 R_1 R_2}{(R_2 - R_1)}$$

For dark fringes:
$$2d + \frac{\lambda}{2} = (2m+1)\frac{\lambda}{2} \Rightarrow d = m\lambda$$

$$x_m = \sqrt{\frac{R_1 R_2 d}{(R_2 - R_1)}} = \sqrt{\frac{R_1 R_2 m\lambda}{(R_2 - R_1)}} \tag{12-22}$$

4. As shown in Fig. 12-10, an initially evacuated tube is placed in one arm of a Michelson interferometer. As a gas is slowly added to the tube, 48 fringes are seen to cross the telescope crosshairs. The length of the tube is 5.00 cm and light source is a mercury-vapor lamp (546 nm). (1) How long the additional optical path length travelled by the light beam is added in this arm? (2) What is the index of refraction of the final quantity of gas within the

tube?

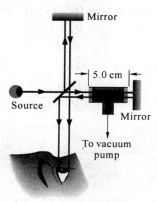

Fig. 12-10　The illustration for Example 4

Solution:

(1) The length of the tube is denoted by l, and the index of refraction of the gas is denoted by n. So the additional optical path length added in this arm is

$$\Delta\delta = 2(n-1)l \qquad (12\text{-}23)$$

(2) As 48 fringes are seen to cross the telescope crosshairs,

$$\Delta\delta = 2(n-1)l = 48\lambda \qquad (12\text{-}24)$$

$$n = \frac{48\lambda}{2l} + 1 = \frac{48 \times 546 \times 10^{-9}}{2 \times 5.00 \times 10^{-2}} + 1 = 1.000,262 \qquad (12\text{-}25)$$

Questions and Problems

1. A double-slit apparatus is used to observe an interference pattern projected onto a screen from a stationary light source. If the light source is instead moved toward the double slits at constant speed, what will be observed on the screen? _____.

　　(A) The interference pattern will remain unchanged

　　(B) The fringes will move farther apart to a new fixed distance

　　(C) The fringes will move closer together to a new fixed distance

　　(D) The fringes will continue to move father and farther apart as the source is brought closer

2. A physics instructor slips a thin sheet of a transparent material with an index of refraction only slightly larger than air over one of the slits in a double-slit demonstration. What will happen to the interference pattern on the screen? _____.

　　(A) There will be no change

　　(B) The fringes will spread further apart

　　(C) The fringes will move closer together

　　(D) The fringes will shift position, but not change spacing

3. Three coherent, equal intensity light rays arrive at a point P on a screen to produce

an interference minimum of zero intensity. If any two of the rays are blocked, the intensity of the light at P is I_1. What is the intensity of the light at P if only one of the rays is blocked? _____.

(A) 0 (B) $I_1/2$ (C) I_1
(D) $2I_1$ (E) $4I_1$

4. 550 nm light strikes a thin film normal to the surface, all of the light is transmitted and none is reflected. A second ray of light of wavelength λ strikes the same thin film at a very small angle to the normal, all of the light is also transmitted and none is reflected. What can be concluded about λ? _____.

(A) $\lambda > 550$ nm (B) $\lambda = 550$ nm (C) $\lambda < 550$ nm

(D) Nothing can be concluded without knowledge of the indices of refraction of the various substances in the problem

5. In Fig. 12-11, two pulse of light are sent through layers of plastic with indices of refraction indicated and with thickness of either L or $2L$. (1) Pulse _____ travels in less time. (2) The difference in the travel time is _____.

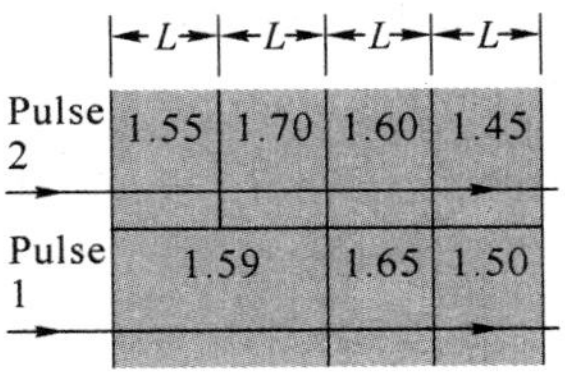

Fig. 12-11 Problem 5

6. In Fig. 12-12, light wave W_1 reflects once from a reflecting surface while light wave W_2 reflects twice from that surface and once from a reflecting silver at distance L from the mirror. The waves are initially in phase and have a wavelength of 620 nm. Neglect the slight tilt of the rays. (1) The value of L must at least _____ in order for the reflected waves to be exactly out of phase. (2) If the silver is moved a length of _____ again, the waves are exactly out of phase again.

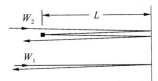

Fig. 12-12 Problem 6

7. The Fig. 12-13 shows four situations in which light reflects perpendicularly from a thin film of thickness L, with indices of refraction as given. For situations _____ and _____, the reflection at the film interfaces cause a zero phase difference for the two reflected rays. For situations _____ and _____, the film will be dark if the path length difference $2L$ causes a phase difference of 0.5 wavelength.

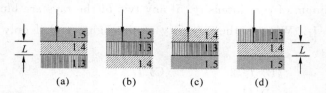

Fig. 12-13 Problem 7

8. Three light waves combine at a certain point where electric field components are
$$E_1 = E_0 \cos \omega t$$
$$E_2 = E_0 \cos(\omega t + 60°)$$
$$E_3 = E_0 \cos(\omega t - 30°)$$
Their resultant component $E(t)$ at that point is _____.

9. Fig. 12-14 shows a transparent plastic block with a thin wedge of air at the right. A broad beam of red light, with wavelength $\lambda = 632.8$ nm, is directed downward through the top of the block. Some of light is reflected back up from the top and bottom surface, which acts as a thin film of air with a thickness that varies uniformly and gradually from L_L at the left-hand end to L_R at the right-hand end. An observer looking down on the block sees an interference pattern consisting of six dark fringes and five bright red fringes along the wedge. What is the change in thickness $\Delta L (= L_L - L_R)$ along the wedge?

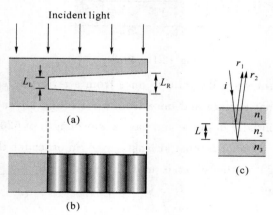

Fig. 12-14 Problem 9

10. Show that (1) in Newton's ring experiment, the difference in radius between adjacent bright ring (maximum) is given by
$$\Delta r = r_{m+1} - r_m \approx \frac{1}{2}\sqrt{\lambda R/m} \quad (m \gg 1)$$
where λ is the wavelength of the light, R is the radius of the lens, and m is the order of the maximum. (2) Now show that the area between adjacent bright rings is given by
$$A = \pi \lambda R \quad (m \gg 1)$$

11. In Fig. 12-15, a microwave transmitter at height a above the water level of a wide lake transmits microwaves of wavelength λ toward a receiver on the opposite shore, a

distance x above the water level. The microwaves reflecting from the water interfere with the microwave arriving directly from the transmitter. Assuming that the lake width D is much greater than a and x, and that $\lambda \geqslant a$, at what values of x is the signal at the receiver maximum?

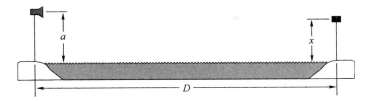

Fig. 12-15　Problem 11

12. If mirror M_2 in a Michelson interferometer is moved through 0.233 mm, a shift of 792 fringes occurs. What is the wavelength of the light producing the fringe pattern?

Chapter 13   Diffraction of Light

Review of the Contents

1. Phenomena of diffraction of light

Diffraction is the deviation of light from a straight-line path when the light passes through an aperture or around an obstacle. The phenomena of Poisson spot and the diffraction pattern produced when a light passes through a single-slit are the two examples of diffraction of light(Fig. 13-1).

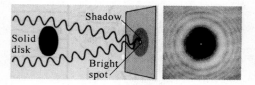

Fig. 13-1   Poisson spot

If either the point source or the viewing screen is relatively close to the barrier, the diffraction is described as near-field diffraction or Fresnel diffraction. If the source, barrier, and screen are far enough away that all the lines from the source to the barrier can be considered parallel and all the lines from the barrier to the pattern can be considered parallel, the phenomenon is called far-field diffraction or Fraunhofer diffraction.

2. Huygens-Fresnel principle

Huygens-Fresnel principle: the amplitude of the wave at any point beyond is the superposition of all these wavelets by taking into account their amplitudes and phases. As shown in Fig. 13-2, if a coherent lightwave falls on a opaque barrier with an open hole, the wave disturbance $E(P)$ at point P on the viewing screen is

$$E(P) = \iint_A \underbrace{K(\theta)}_{\text{Obliquity factor}} \underbrace{\frac{E(A)}{r}\cos\left(\omega t - \frac{2\pi}{\lambda}r\right)}_{\text{Spherical wavelet}} dA \tag{13-1}$$

where $E(A)$ is the amplitude of the wavelet at point at the surface of the open hole, r is the distance between A and P, λ is the wavelength of the incident lightwave. θ is obliquity angle of ray AP deviating from the normal direction of the hole surface.

3. Single-slit Fraunhofer diffraction

Fresnel half-period zone treatment: Every point at the wavefront of the slit can be

Chapter 13 Diffraction of Light

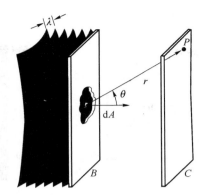

Fig. 13-2 The illustration of Huygens-Fresnel principle

considered as the source of secondary wavelet (as shown in Fig. 13-3). All rays that can focus on point P on the screen are a series of parallel lines which make the angle θ with horizontal line. A series of parallel planes with the equal spacing $\lambda/2$, which are perpendicular to the rays, divide the slit into a series of narrow strips. These strips are called Fresnel half-period zones. The contributions at P from adjacent zones are half-cycle out of phase and tend to cancel.

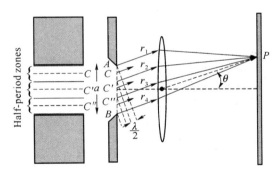

Fig. 13-3 Fresnel half-period zones

See Fig. 13-4, for the case of $\theta=0$, rays that leave the slit parallel to the central horizontal axis are brought to a focus at central point $y=0$, and in phase at this point. They interference constructively and produce a central intensity maximum at $y=0$.

For an angle θ, the slit just can be divided into even number of half-period zones, these zones cancel each other and make the minima at P on the screen.

$$a\sin\theta = \pm 2m\frac{\lambda}{2} = \pm m\lambda \quad (m=1,2,3,\cdots) \tag{13-2}$$

but not at $m=0$ where there is the central strongest maximum.

For an angle θ, the slit just can be divided into odd number of half-period zones, there is always a zone left not to be canceled, and makes the maxima on the screen.

$$a\sin\theta = \pm(2m+1)\frac{\lambda}{2} \quad (m=1,2,\cdots) \tag{13-3}$$

The diffraction pattern of a single-slit is shown in Fig. 13-4.

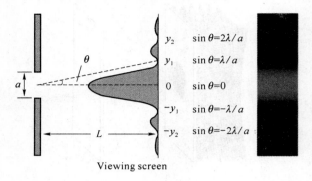

Fig. 13-4 Single-slit diffraction

Phasor treatment: As shown in Fig. 13-5, divide the slit of width a into N strips of width Δy. The phase difference in the rays coming from adjacent strips will be $\Delta\beta = \frac{2\pi}{\lambda}\Delta y \sin\theta$; the total amplitude on the screen at the angle θ will be the sum of separate wave amplitudes due to each strip. Each differs in phase from adjacent one by $\Delta\beta$. So totally

$$\beta = N\Delta\beta = \frac{2\pi}{\lambda}a\sin\theta \tag{13-4}$$

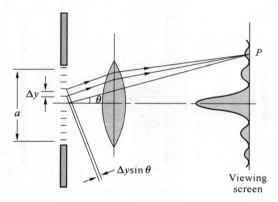

Fig. 13-5 Divide the width a into N strip of width Δy

Using the phasor diagrams in Fig. 13-6, we get

$$E_P = E_0\frac{\sin(\beta/2)}{\beta/2},\ I = I_0\left[\frac{\sin(\beta/2)}{\beta/2}\right]^2 \tag{13-5}$$

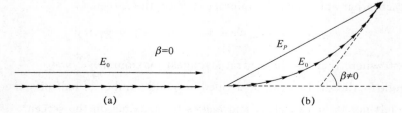

Fig. 13-6 The phasor diagrams when $\beta = 0$ (a), and $\beta \neq 0$ (b)

Chapter 13　Diffraction of Light

The intensity distribution of single-slit diffraction pattern is shown in Fig. 13-7. We can see that the central bright band of a single-slit diffraction pattern is twice as wide as other side bright bands, and shows much brighter than fainter side bands since it holds more than 80% of the power of the whole diffraction pattern. The half-width θ_1 characterizes the angular spread of the diffraction, and is given by

$$\sin\theta_1 \approx \theta_1 = \frac{\lambda}{a} \qquad (13\text{-}6)$$

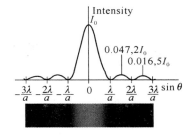

Fig. 13-7　The intensity distribution of single-slit diffraction pattern

4. Circular apertures and resolution of an optical instrument

The diffraction pattern formed by a circular aperture consists of a central bright spot, called Airy disk, surrounded by a series of bright and dark rings, as shown in Fig. 13-8. The angular half-width of Airy disk is given by

$$\sin\theta_a \approx \theta_a = 1.22\frac{\lambda}{D} \qquad (13\text{-}7)$$

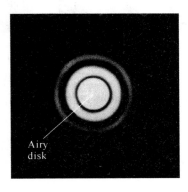

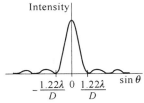

Fig. 13-8　Diffraction of circular aperture

Image formation blurred by diffraction: Because light travels as a wave, light from a point source passing through a lens, which is a kind of the circular aperture, is spread out into a diffraction pattern. When a lens forms the image of a point object, the image is actually a tiny diffraction pattern rather than a point. Thus an image would be blurred even if aberrations were absent.

Rayleigh criterion: Two images are just resolvable when center of diffraction Airy disk of one image is directly over the first minimum in the diffraction pattern of the other, as shown in Fig. 13-9. Two objects can be considered just resolvable if they are separated by an angle

$$\theta_{\min} = \theta_a = 1.22 \frac{\lambda}{D} \tag{13-8}$$

The resolving power of a lens is given by

$$R = \frac{1}{\theta_{\min}} = \frac{D}{1.22\lambda} \tag{13-9}$$

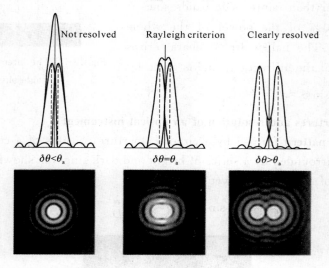

Fig. 13-9 Image resolving judgment by Rayleigh criterion

5. The diffraction grating

As shown in Fig. 13-10, a large number of equally spaced parallel slits is called a diffraction grating. The intensity distribution of the diffraction pattern of a diffraction grating is given by

$$I = I_0 \left(\frac{\sin \beta/2}{\beta/2}\right)^2 \left(\frac{\sin N\gamma/2}{\sin \gamma/2}\right)^2 \tag{13-10}$$

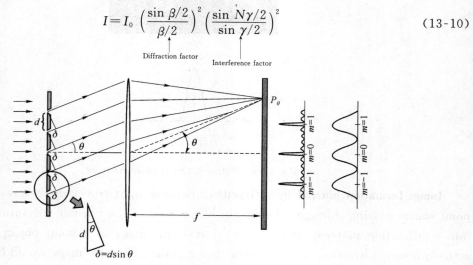

Fig. 13-10 The illustration of a diffraction grating

where I_0 is the intensity at the central position of a single-slit diffraction pattern. β, γ factors are given by

$$\beta = \frac{2\pi}{\lambda} a \sin \theta \tag{13-11}$$

Chapter 13 Diffraction of Light

which represents the total phase difference in the rays coming from two edges of a single slit.

$$\gamma = \frac{2\pi}{\lambda} d \sin \theta \tag{13-12}$$

which represents the phase difference between rays from any pair of adjacent slits.

Principal maxima locate where $d \sin \theta = \pm m\lambda$ $(m = 0, \pm 1, \pm 2, \cdots)$ (13-13)

Half-width of the principal maxima: $\Delta \theta_m = \dfrac{\lambda}{Nd \cos \theta_m}$ (13-14)

Minima locate where $d \sin \theta = \pm \dfrac{m'}{N} \lambda$ $(m' \neq 0, N, 2N, \cdots)$ (13-15)

As shown in Fig. 13-11, there are $N-1$ minima between two adjacent principal maxima, and $N-2$ secondary maxima.

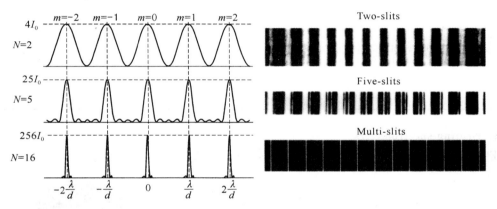

Fig. 13-11 The intensity distributions and diffraction patterns for 2, 5, 16 slits gratings

Missing order: If the mth-order interference maximum coincides with the nth-order diffraction minimum, with the result that $\dfrac{d}{a} = \dfrac{m}{n}$, the mth-order principal maximum will be missed (Fig. 13-12).

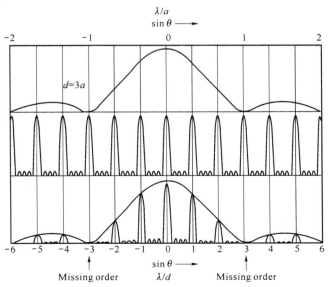

Fig. 13-12 The illustration of missing order

Highest order of principal maximum: Since the diffraction pattern distribution is limited in the angular range from $-90°$ to $90°$, the highest order of bright fringe we can see satisfies $d|\sin\theta|=m_{max}\lambda$. Therefore $m_{max}=\pm\left[\dfrac{d}{\lambda}\right]$.

All the features of a grating are summarized in Table 13-1.

Table 13-1 Summary of a grating

	Principal maxima				Minima			Secondary maxima	
	Location	Intensity	Half-width $\Delta\theta$	Missing order	Location	Number	Location	Number	
d	$\sin\theta=m\dfrac{\lambda}{d}$	×	$\Delta\theta=\dfrac{\lambda}{Nd\cos\theta}$	$\dfrac{d}{a}=\dfrac{m}{n}$	$\sin\theta=\dfrac{m'}{N}\dfrac{\lambda}{d}$ $(m'\neq 0,\pm N,\pm 2N,\cdots)$	×	✓	×
a	×	Modulation	×	$\dfrac{d}{a}=\dfrac{m}{n}$	×	×	×	×
N	×	N^2I_0	$\Delta\theta=\dfrac{\lambda}{Nd\cos\theta}$	×	$\sin\theta=\dfrac{m'}{N}\dfrac{\lambda}{d}$ $(m'\neq 0,\pm N,\pm 2N,\cdots)$	$N-1$	✓	$N-2$

6. Grating spectrometers

The configuration of a spectrometer is shown in Fig. 13-13.

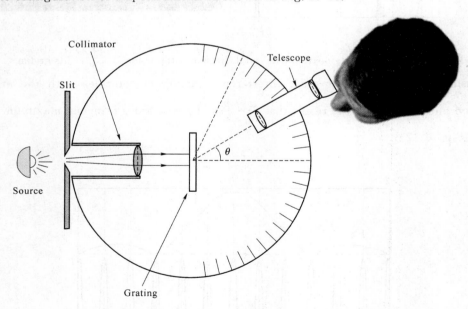

Fig. 13-13 The configuration of a spectrometer

Like a prism, a diffraction grating can be characterized by its dispersion D as

$$D=\frac{\Delta\theta}{\Delta\lambda}=\frac{m}{d\cos\theta} \tag{13-16}$$

Therefore a diffraction grating can be used as the dispersion element in a spectrometer used for spectrum analysis.

Resolving power: For two nearly equal intensity lights with wavelengths λ_1 and λ_2

Chapter 13　Diffraction of Light

between which a diffraction grating can just barely distinguish according to Rayleigh criterion, the resolving power of the grating in the mth-order of diffraction pattern is given by

$$R = \frac{\lambda_{\text{average}}}{\Delta\lambda} = mN \tag{13-17}$$

The spectrum for hydrogen in the visible range measured by a diffraction grating spectrometer is shown in Fig. 13-14.

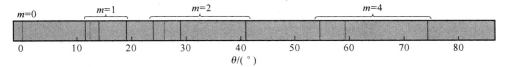

Fig. 13-14　The 0th, 1st, 2nd, and 4th order of the visible emission lines from hydrogen (410.1 nm, 434 nm, 486.1 nm, 656.2 nm), measured by a diffraction grating spectrometer

7. Diffraction of X-rays by crystals

The regular array of atoms in a crystal is a three-degree-dimensional diffraction grating for short-wavelength waves such as X-rays. For analysis purpose, the atoms can be visualized as being arranged in planes with characteristic interplanar spacing d. Diffraction maxima (due to constructive interference) occur if the incident direction of the wave, measured from the surface of these planes, and the wavelength λ of the radiation satisfy Bragg's law(Fig. 13-15):

$$2d\sin\phi = m\lambda \tag{13-18}$$

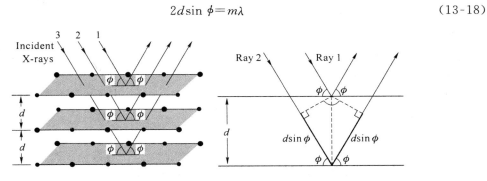

Fig. 13-15　The illustration of Bragg's law

Typical Examples

Problem solving strategy

1. Table 13-1 is a good summary of most of contents of multi-slit diffraction. A multi-slit diffraction can be considered as the combination of the multi-slit interference (corresponds to the interference factor in Eq. (13-10)) and the single-slit diffraction (corresponds to the diffraction factor in Eq. (13-10)). In a diffraction pattern, the effect of multi-slit interference appears as the principal maxima or the fringes in the diffraction pattern, while the effect of single-slit diffraction makes the envelope of the diffraction

pattern or the modulation of the whole intensity distribution.

2. The principal maxima of the multi-slit interference factor take place when the optical path difference is:
$$d\sin\theta = \pm m\lambda \quad (m=0,\pm 1,\pm 2,\cdots)$$
The minima of the single-slit diffraction factor take place when the optical path difference is:
$$a\sin\theta = \pm k\lambda \quad (k=1,2,3,\cdots)$$

3. The number of the bright fringes in the central peak of the diffraction envelope is $2\left[\dfrac{b}{a}\right]+1$. When $\dfrac{d}{a}=\dfrac{m}{n}$, the mth-order bright fringe will be missed. The highest order of bright fringe is $m_{\max}=\pm\left[\dfrac{d}{\lambda}\right]$.

4. The phasor is a helpful and visualized method to analyze the diffraction phenomena. We may use it as possible as we can.

Examples

1. A single silt is illuminated by light of wavelength λ_a and λ_b, which are chosen so that the first diffraction minimum of the λ_a component coincides with the second minimum of the λ_b component. (1) What relationship exists between the two wavelengths? (2) Do any other minima in the two diffraction patterns coincide?

Solution:

(1) The first diffraction minimum of the λ_a component coincides with the second minimum of λ_b component.
$$a\sin\theta = \lambda_a = 2\lambda_b$$
So the relationship exists between the two wavelength is
$$\lambda_a = 2\lambda_b$$

(2) From $a\sin\theta = \lambda_a = 2\lambda_b$, we have
$$a\sin\theta_m = m\lambda_a = 2m\lambda_b \quad (m=1,2,\cdots)$$
which means that mth-order minima of the λ_a component coincides with $2m$th-order minima of the λ_b component.

2. A 0.10 mm wide slit is illuminated by light of wavelength 589 nm. Consider a point P on a viewing screen on which the diffraction pattern of the slit is viewed; the point is at $30°$ from the central axis of the slit. What is the phase difference between the Huygens wavelets arriving at P from the top and the midpoint of the slit?

Solution:

In Fig. 13-16 we get the path difference between the Huygens wavelets arriving at P from the top and the midpoint of the slit is
$$\delta = \frac{a}{2}\sin\theta$$

Chapter 13 Diffraction of Light

So the corresponding phase difference is

$$\Delta\varphi = \frac{2\pi}{\lambda}\frac{a}{2}\sin\theta = \frac{\pi}{\lambda}a\sin\theta$$

$$= \frac{\pi}{589\times 10^{-9}}\times 0.10\times 10^{-3}\sin 30°$$

$$= 84.89\pi = 266.68 \text{ rad}$$

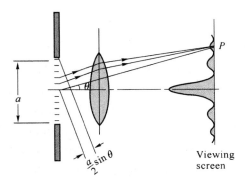

Fig. 13-16 The illustration for Example 2

3. The full width at half-maximum (FWHM) of a central diffraction maximum is defined as the angle between the two points in the pattern where the intensity is one-half that at the center of the pattern (Fig. 13-17). (1) Show that the intensity drops to one-half the maximum value when $\sin^2\beta = \beta^2/2$ $\left(\beta = \frac{\pi}{\lambda}a\sin\theta\right)$. (2) Verify that $\beta = 1.39$ rad (about 80°) is a solution to the transcendental equation of (1). (3) Show that the FWHM is $\Delta\theta = 2\sin^{-1}(0.443\lambda/a)$, where a is the slit width. (4) Calculate the FWHM of the central maximum for slits whose width are 1.0, 5.0, and 10 wavelengths.

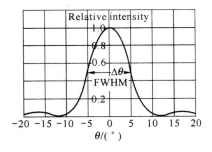

Fig. 13-17 The illustration for Example 3

Solution:

(1) The intensity of diffraction of a single-slit is

$$I = I_0 \frac{\sin^2\beta}{\beta^2}$$

When the intensity drops to one-half the maximum value, we get

$$\frac{I_0}{2} = I_0 \frac{\sin^2\beta}{\beta^2} \Rightarrow \sin^2\beta = \beta^2/2$$

(2) When $\beta=1.39$, $\sin^2\beta=\sin^2 1.39 \text{ rad}=0.968$, $\beta^2/2=1.39^2/2=0.966$. To be sure that 1.39 rad is closer to the correct value.

(3) Since $\beta=(\pi a/\lambda)\sin\theta$, now $\beta/\pi=1.39/\pi=0.442$, so

$$\theta=\arcsin\left(\frac{0.442\lambda}{a}\right)$$

The angular separation of the two points of half intensity, one on either side of the center of the diffraction pattern, is

$$\Delta\theta=2\theta=2\arcsin\left(\frac{0.442\lambda}{a}\right)$$

(4) For $a/\lambda=1.0$, $\Delta\theta=2\theta=2\arcsin(0.442/1.0)=0.916 \text{ rad}=52.5°$.
For $a/\lambda=5.0$, $\Delta\theta=2\theta=2\arcsin(0.442/5.0)=0.177 \text{ rad}=10.1°$.
For $a/\lambda=10$, $\Delta\theta=2\theta=2\arcsin(0.442/10)=0.088, 4 \text{ rad}=5.06°$.

4. Light of wavelength 440 nm passes through a double slit, yielding a diffraction pattern whose graph of intensity versus angular position θ is shown in Fig. 13-18. Calculate (1) the slit width and (2) the slit separation. (3) Verify the displayed intensities of $m=1$ and $m=2$ interference fringes.

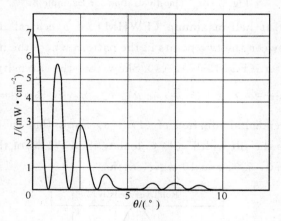

Fig. 13-18   The illustration for Example 4

Solution:
(1) The first minimum of the diffraction pattern is at $5.00°$, so

$$a=\frac{\lambda}{\sin\theta}=\frac{0.440 \text{ }\mu\text{m}}{\sin 5.00°}=5.05 \text{ }\mu\text{m}$$

(2) Since the fourth bright fringes is missing, $d=4a=4\times 5.05 \text{ }\mu\text{m}=20.2 \text{ }\mu\text{m}$.

(3) For the $m=1$ bright fringe,

$$\beta_1=\frac{\pi}{\lambda}a\sin\theta_1=\frac{\pi\times 5.05 \text{ }\mu\text{m}\times\sin 1.25°}{0.440 \text{ }\mu\text{m}}=0.787 \text{ rad}$$

The intensity of the $m=1$ fringe is

$$I_1=I_0\left(\frac{\sin\beta_1}{\beta_1}\right)^2=(7.0 \text{ mW/cm}^2)\left(\frac{\sin 0.787}{0.787}\right)^2=5.7 \text{ mW/cm}^2$$

which agrees with Fig. 13-18.

Similarly, for $m=2$ fringe,

$$\beta_2 = \frac{\pi}{\lambda} a \sin \theta_2 = \frac{\pi \times 5.05 \ \mu m \times \sin 2.50°}{0.440 \ \mu m} = 1.57 \text{ rad}$$

$$I_1 = I_0 \left(\frac{\sin \beta_1}{\beta_1}\right)^2 = (7.0 \text{ mW/cm}^2) \left(\frac{\sin 1.57}{1.57}\right)^2 = 2.8 \text{ mW/cm}^2$$

which also agrees with Fig. 13-18.

5. A grating has 600 rulings/mm and is 5.0 mm wide. (1) What is the smallest wavelength interval it can resolve in the third order at $\lambda = 500$ nm? (2) How many higher orders of maxima can be seen?

Solution:

(1) The total number of slits is

$$N = \frac{\text{Total width of the grating}}{\text{Width of slit per each ruling}} = 5.0 \text{ mm} \times 600 \text{ rulings/mm} = 3,000$$

$$\Delta \lambda_{min} = \frac{\lambda_{average}}{mN} = \frac{500 \text{ nm}}{3 \times 3,000} = 0.005,6 \text{ nm}$$

(2) $m_{max} = \pm \left[\dfrac{d}{\lambda}\right] = \pm \left[\dfrac{1/(600 \text{ rulings}/10^6 \text{ nm})}{500 \text{ nm}}\right] = \pm 3$

There will be ± 2, ± 3 higher order maxima can be seen.

6. Fig. 13-19 is a graph of intensity versus angular position θ for the diffraction of an X-ray beam by a crystal. The beam consists of two wavelengths, and spacing between the reflecting planes is 0.94 nm. What are the two wavelengths?

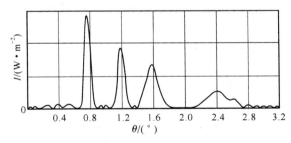

Fig. 13-19 The illustration for Example 6

Solution:

As shown in Fig. 13-20, from Bragg's law, $2d \sin \phi = m\lambda$, and we find $\theta = 2\phi$.

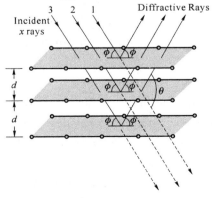

Fig. 13-20 The illustration for Example 6

From Fig. 13-19, we can get the conclusions that
For λ_1,
$$m=1, \theta_1=0.72 \text{ rad}$$
$$m=2, \theta_1=1.58 \text{ rad}$$
So, we get
$$\lambda_1=0.66 \text{ nm}$$
For λ_2,
$$m=1, \theta_2=1.15 \text{ rad}$$
$$m=2, \theta_2=2.50 \text{ rad}$$
So, we get
$$\lambda_2=1.02 \text{ nm}$$

Questions and Problems

1. Which of these statements is most correct? _____.
   (A) Diffraction occurs only for transverse waves
   (B) Diffraction is proof that light can behave like a wave
   (C) Diffraction explains rainbows
   (D) Red sunsets are the diffraction phenomena

2. Two colors of light, red and violet, are allowed to pass through a single slit. It is observed that for a certain slit width, a, the second maximum of violet and the first maximum of the red light overlap. The slit width is then decreased. What happens to the overlapping violet and red maxima? _____.
   (A) The maxima separate, with the angle of the red maximum increasing faster than the violet
   (B) The maxima separate, with the angle of the violet maximum increasing faster than the red
   (C) The angle of the red maximum increases while the angle of the violet decreases
   (D) The angle for both maxima increases so they continue to overlap

3. The intensity of the first maximum beyond the central maximum is approximately what percentage of the intensity of central maximum? _____.
   (A) 50%　　　　　　　　　　　(B) 25%
   (C) 5%　　　　　　　　　　　　(D) It depends on the width of the slit

4. Diffraction pattern minima are located at the angles θ_m, where m is a positive nonzero nonnegative integer and _____.
   (A) $a\sin\theta_m = m\lambda$　　　(B) $a\sin\theta_m > m\lambda$　　　(C) $a\sin\theta_m < m\lambda$

5. Consider diffraction from a circular aperture of diameter D. The first minimum will occur at $D\sin\theta = k\lambda$, where k is a constant. What value would you expect for k? _____.
   (A) $k<1$　　　(B) $k=1$　　　(C) $1<k<1.22$
   (D) $k=1.22$　　(E) $k>1.22$

6. In a single-silt diffraction experiment, the top and bottom rays through the slit arrive at a certain point on the viewing screen with a path length difference of 4.0 wavelengths. In

Chapter 13 Diffraction of Light

a phasor representation, the chain of phasors makes _____ (how many) overlapping circles.

7. Fig. 13-21(a) and (b) represent the phasor diagrams for two points of a diffraction pattern that are on opposite sides of a certain diffraction maximum. Which maximum is it? _____.

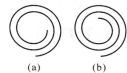

Fig. 13-21 Problem 7

8. Two wavelengths, 650 nm and 430 nm, are used separately in a single-slit diffraction experiment. Fig. 13-22 shows the results as graphs of intensity I versus angle θ for the two diffraction patterns. If both wavelengths are then used simultaneously, what color will be seen in the combined diffraction pattern at (1) angle A and (2) angle B? _____.

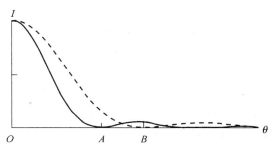

Fig. 13-22 Problem 8

9. Suppose that you can barely resolve two red dots, owing to diffraction by the pupil of your eye. If we increase the general illumination around you so that the pupil decrease in diameter, does the resolvability of dots improve or diminish? Consider only diffraction. _____.

10. Fig. 13-23 shows lines of different orders produced by a diffraction grating in monochromatic red light. (1) Is the center of the pattern to the left or right? (2) If we switch to monochromatic green light, will the half-widths of the lines then produced in the same orders be greater than, less than, or the same as the half-widths of the lines shown? _____.

Fig. 13-23 Problem 10

11. Fig. 13-24 shows the bright fringes that lie within the central diffraction envelope in two double-slit diffraction experiments using the same wavelength of light. Are (1) the slit width a, (2) the slit separation d, and (3) the ratio d/a in experiment B greater than, less than, or the same as those in experiment A? _____.

Fig. 13-24 Problem 11

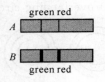

Fig. 13-25   Problem 12

12. Fig. 13-25 shows the lines produced by diffraction gratings A and B using light of the same wavelength; the lines are of the same order and appear at the same angles θ. (1) Which grating has the greater number of rulings? (2) Is the center of the diffraction pattern to the left or to the right? _____.

13. Fig. 13-26 shows lines of two orders produced by a single diffraction grating using light of two wavelengths, both in the red region of the spectrum. (1) Which lines, the left pair or right pair, are in the order with greater m? (2) Is the center of the diffraction pattern to the left or to the right? _____.

Fig. 13-26   Problem 13

14. Fig. 13-27 shows the interference pattern (the intensity distribution vs. angular location without considering the single-slit diffraction factor) produced by an equally spaced N-slit grating. The marks a, b and c on the horizontal axis denote three of the minima between zero-order and the first-order principal maxima. (1) What is the number N of slits for this grating? (2) Draw the appropriate phasor diagrams corresponding to these three minima. (3) Label the locations on the horizontal axis for these three minima using the quantities λ (the wavelength), d (grating spacing) and N (the number of slits).

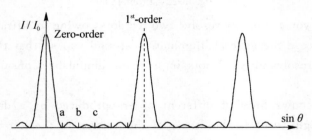

Fig. 13-27   Problem 14

15. Light of wavelength 600 nm is incident normally on a diffraction grating. Two adjacent maxima occur at angles given by $\sin\theta = 0.2$ and $\sin\theta = 0.3$. The fourth-order maximum is missing. (1) What is the separation between adjacent slits? (2) What is the smallest slit width this grating can have? (3) Name all orders that appear on the viewing screen, consistent with the answers to parts (1) and (2).

Chapter 14 Polarization of Light

Review of the Contents

1. States of polarization (SOP)

For a plane electromagnetic wave, the **E** and **B** are perpendicular to each other and to the direction of travel, which is the basic picture of a transverse wave (Fig. 14-1). By convention, we call the **E** vector as the light vector, and define the direction of the **E** vector as to be the **direction of polarization**. The plane determined by the **E** vector and the direction of propagation of the wave is called the **plane of polarization** of the wave.

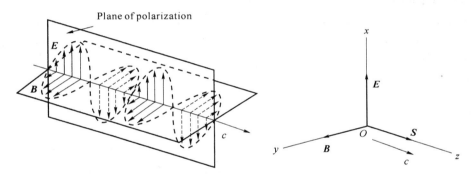

Fig. 14-1 Electric and magnetic field strengths in an electromagnetic wave

E and **B** are at right angles to each other. The entire pattern moves in a direction perpendicular to both **E** and **B**, which is the direction of Poynting vector $\boldsymbol{S}=\dfrac{1}{\mu_0}\boldsymbol{E}\times\boldsymbol{B}$, the energy current density of electromagnetic waves.

There are five kinds of SOPs of light in the view of University Physics, linearly polarized, circularly polarized, elliptically polarized, unpolarized, and partially polarized, respectively.

If z axis is to be the direction of travel, any polarized wave vector **E** can be considered to be a vector sum of two orthogonal components as

$$\boldsymbol{E}(z,t)=\boldsymbol{E}_x(z,t)+\boldsymbol{E}_y(z,t)$$

where

$$\begin{aligned}&E_x(z,t)=A_x\cos(\omega t-kz+\phi_x)\\&E_y(z,t)=A_y\cos(\omega t-kz+\phi_y)\end{aligned} \quad \Delta\phi=\phi_y-\phi_x \tag{14-1}$$

Linearly polarized light: A light is said to be linearly polarized when $\Delta\phi=0$ or π as shown in Fig. 14-2. $\Delta\phi=0$ corresponds to the light linearly polarized in Ⅰ-Ⅲ quadrant, and

$\Delta\phi=\pi$ corresponds to the light linearly polarized in II-IV quadrant.

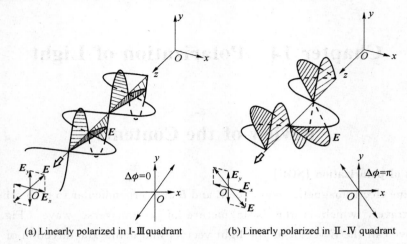

(a) Linearly polarized in I-III quadrant (b) Linearly polarized in II-IV quadrant

Fig. 14-2 Linearly polarized light

Circularly polarized light: A light is said to be circular polarized if the electric field $\boldsymbol{E}$ rotates uniformly in the plane perpendicular to the propagation direction at wave frequency. The circular polarized light is said to be right-circular if rotating clockwise when $\Delta\phi=\pi/2$, and to be left-circular if rotating counterclockwise when $\Delta\phi=-\pi/2$ (Fig. 14-3).

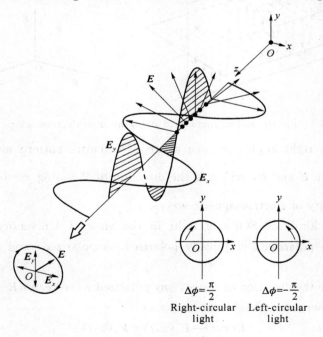

Fig. 14-3 Circularly polarized light

Elliptically polarized light: A light is said to be elliptically polarized (or simply elliptical light) if the end of electric field vector $\boldsymbol{E}$ draws an ellipse in the plane perpendicular to the propagation direction at wave frequency. The elliptically polarized light is said to be right-handed when rotating clockwise, and to be left-handed when rotating counterclockwise

Chapter 14 Polarization of Light

(Fig. 14-4).

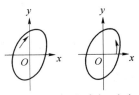

Right-handed   Left-handed
elliptical light elliptical light

Fig. 14-4 Elliptically polarized light

Unpolarized light or natural light: A light is said to be unpolarized if the direction of **E** vector change randomly with time. We can mathematically represent natural light in terms of two arbitrary, incoherent, orthogonal, linearly polarized waves with equal amplitude but random phase difference between them, as shown in Fig. 14-5.

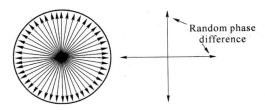

Fig. 14-5 Unpolarized light

Partially polarized light: A light is generally neither completely polarized nor completely unpolarized; both cases are extremes. More often, the electric field vectors **E** varies in a way that is neither totally regular nor totally irregular, and such light is said to be partially polarized light. We can mathematically represent partially polarized light in terms of mixture of specific amount of natural light and polarized light, as shown in Fig. 14-6.

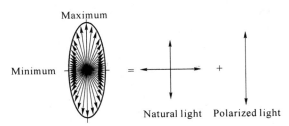

Fig. 14-6 Partially polarized light

2. Polarizer

A linearly polarized light beam can be produced by an unpolarized light passing through a polarizer. The dichroic crystal (shown in Fig. 14-7), which has an optic axis called polarizing axis, possesses a property that: the electric-field component of an incident light that is perpendicular to the polarizing axis is strongly absorbed than another orthogonal component. When a natural light passes through such dichroic crystal, it will be predominantly polarized in the direction of crystal's polarizing axis. If the crystal is thick

· 155 ·

enough, the output light will be linearly polarized in polarizing axis (also called transmission axis).

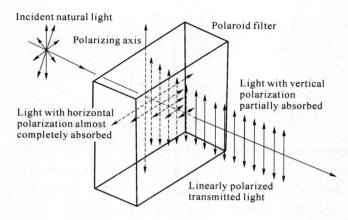

Fig. 14-7　A dichroic crystal

Polaroid sheet: The Polaroid sheet is an artificial device that has dichroism, which can be made in larger size. A polaroid sheet consists of complicated long molecules chains, in a large transparent film plate, arranged parallel to one another.

Degree of polarization: The degree of polarization is defined as the fraction of polarization in total light intensity

$$P = \left(\frac{I_{\text{polarized}}}{I_{\text{total}}}\right) = \frac{I_{\max} - I_{\min}}{I_{\max} + I_{\min}} \tag{14-2}$$

where $I_{\max}$ and $I_{\min}$ are the maximum and minimum intensities that are obtained when the light passes through a polarizer that is slowly rotated.

Malus's law: As shown in Fig. 14-8, if a linear light strikes a polaroid whose axis at an angle θ to the incident polarization direction, the perpendicular component amplitude vanished, and the parallel amplitude $E_0 \cos\theta$ will pass through the polaroid. The intensity of the linear light transmitted by a polarizer is:

$$I = I_0 \cos^2\theta \tag{14-3}$$

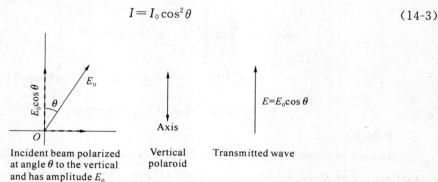

Fig. 14-8　Malus's law

3. Polarization by reflection

As shown in Fig. 14-9, unpolarized light can be polarized, either partially or totally, by

Chapter 14 Polarization of Light

reflection. Light reflected from the smooth surface of water in a lake is partially polarized parallel to the surface.

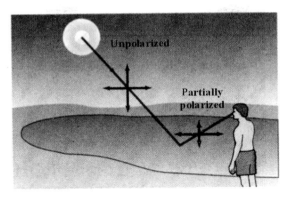

Fig. 14-9 Light reflected from the surface of water in a lake is partially polarized parallel to the surface

Brewster's angle: The degree of polarization of reflected light is varied with incident angle. At one particular angle of incidence, called polarizing angle θ_p, or Brewster's angle, the reflected light is totally polarized, with its plane of polarization perpendicular to the plane of incidence. At the Brewster's angle, the reflected ray and the refracted ray are perpendicular to each other (Fig. 14-10).

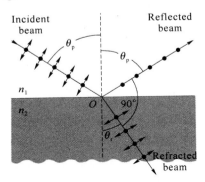

Fig. 14-10 At θ_p the reflected light is linearly polarized parallel to the surface, and $\theta_p + \theta_r = 90°$, where θ_r is the refraction angle
(The dots represents vibration perpendicular to the page)

$$\tan \theta_p = \frac{n_2}{n_1} \tag{14-4}$$

4. Double refraction or birefringence

Optical isotropic and anisotropic: Many crystalline substances are anisotropic, whose the index of refraction is dependent on the direction of propagation in the medium and the state of polarization of the light. Fig. 14-11, in which a polished calcite ($CaCO_3$) is laid over a printed pattern, shows the anisotropy of this of crystalline substance; the image appears double. The phenomena of the single beam splitting into two at the crystal surface, or the phenomena of "double bending" of a beam transmitted through the crystals, is called double refraction or birefringence.

Fig. 14-11 Double image formed by a calcite crystal

O-ray and e-ray: As shown in Fig. 14-12, when a single unpolarized light beam splits into two beams at the crystal surface, the beam that obeys the ordinary Snell's law of refraction is called the ordinary ray or o-ray, and the other beam that does not obey Snell's law is called the extraordinary ray or e-ray. It is found that the o-ray and e-ray images are all linearly polarized.

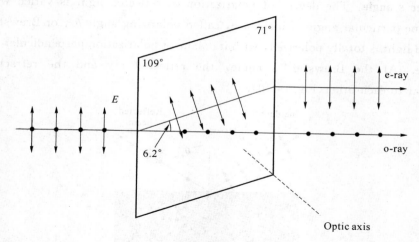

Fig. 14-12 Unpolarized light falling on a calcite crystal splits into two components, the o-ray(which follows Snell's law of refraction) and e-ray (which does not follow Snell's law). The two refracted rays have perpendicular polarizations

Optic axis: There is a characteristic direction in the anisotropic crystal, in whose direction there is no birefringence phenomenon. The optic axis is actually a direction and not merely a single line. For a calcite crystal in its cleavage form shown in Fig. 14-13, there are two blunt corners where the surface planes meet to form three obtuse angles. The line passing through the vertex of either of the blunt corners, oriented so that it makes equal angles with each face and each edge, corresponds to the optic axis.

A plane that contains both the optic axis and the o-ray is called the principal plane of o-ray. The o-ray beam is linearly polarized perpendicular to its principal plane. A plane that contains both the optic axis and the e-ray is called the principal plane of e-ray. The e-ray

Chapter 14 Polarization of Light

beam is linearly polarized parallel to its principal plane. In particular, if the incident light beam is in the same plane with normal of the surface of a crystal, the principal planes of both o-ray and e-ray converge to one plane, and the polarization directions of o-ray and e-ray are at right angles to each other.

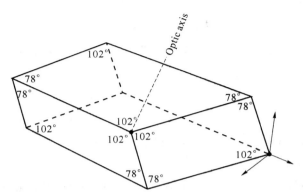

Fig. 14-13   The optic axis for a calcite crystal

Huygens's treatment for birefringence: As shown in Fig. 14-14, a point source embedded in a calcite emits two Huygens wave surfaces. The o-wave surface is a sphere, because the medium is isotropic for o-waves. The e-wave surface is an ellipsoid of revolution about the optic axis. We can describe o-wave and e-wave as follows: (1) The o-wave travels in the crystal with the same speed v_o in all direction. In other words, the crystal has, for o-wave, a single index of refraction n_o. (2) The e-wave travels in the crystal with a speed that varies with direction from v_o to v_e. In other words, the index of refraction, defined as c/v, varies with direction from n_o to n_e.

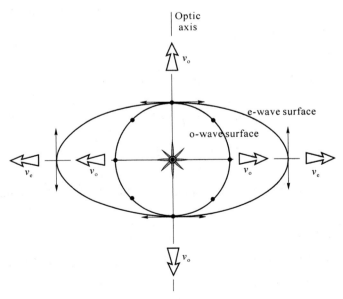

Fig. 14-14   Huygens wave surfaces produced by a point source embedded in a calcite crystal

Typical Examples

Problem solving strategy

1. How to generate a linearly polarized light beam:

A linearly polarized light beam can be generated (1) by a Polaroid sheet; (2) by a glass plate at the Brewster's angle; (3) by separation of o-ray or e-ray passing through a birefringence crystal.

2. Any light beam with any state of polarization is launched into a Polaroid sheet, the emerging light beam is linearly polarized, and the output intensity of the beam is related to the input by Malus' law.

3. Huygens wavelet theory is crucial to treat double refraction using o-wave surface and e-wave surface.

Examples

1. Show that, for an incident partially polarized light beam, the light transmitted by a polarizer, whose axis makes an angle ϕ to the direction in which I_{max} is obtained, has intensity

$$\frac{1+P\cos 2\phi}{1+P} I_{max}$$

where P is the degree of polarization of the partially polarized light beam. I_{max} and I_{min} are the maximum and minimum intensities that are obtained when the light passes through a polarizer that are slowly rotated.

Solution:

From the definition of the degree of polarization we have

$$P = \frac{I_{max} - I_{min}}{I_{max} + I_{min}}$$

we have

$$I_{min} = \frac{1-P}{1+P} I_{max}$$

As shown in Fig. 14-15, the components of the electric field amplitudes along the polarizing axis are

$$E_{max}\cos\phi \text{ and } E_{min}\sin\phi$$

The intensity transmitted by the polarizer is the addition of $(E_{max}\cos\phi)^2$ and $(E_{min}\sin\phi)^2$.

$$I = E_{max}^2 \cos^2\phi + E_{max}^2 \sin^2\phi = I_{max}\cos^2\phi + I_{min}\sin^2\phi$$

$$= I_{max}\cos^2\phi + \frac{1-P}{1+P} I_{max}\sin^2\phi = \frac{1+P\cos 2\phi}{1+P} I_{max}$$

Chapter 14   Polarization of Light

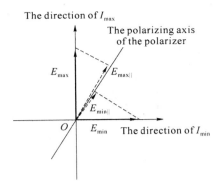

Fig. 14-15   The illustration for Example 1

2. An unpolarized light beam is normally passed through two stacked polarizing sheets. Find the angle between the two polarizing sheets, when (1) the intensity of transmitted light is 1/7 of the incident, or (2) the intensity of transmitted light is 1/4 of that achieves its maximum.

Solution:

Let the intensity of the unpolarized light is I_0.

(1) The intensity that is passed through the first polarizing sheet is

$$I_1 = \frac{1}{2} I_0$$

The intensity that is passed through the second polarizing sheet is

$$I_2 = I_1 \cos^2 \theta = \frac{1}{2} I_0 \cos^2 \theta = \frac{1}{7} I_0$$

$$\cos \theta = \sqrt{\frac{I_2}{I_0/2}} = \sqrt{\frac{2}{7}}$$

$$\theta = 57°41'$$

(2) When the axis of two polarizing sheets is parallel to each other, the transmitted light achieves its maximum intensity, which is equal to $I_0/2$. Assume the angle between the two sheets is θ, then the intensity that is passed through the two sheets is

$$I_2 = I_1 \cos^2 \theta = \frac{1}{2} I_0 \cos^2 \theta = \frac{1}{4} \cdot \frac{I_0}{2}$$

$$\cos \theta = \sqrt{\frac{1}{4}} = \frac{1}{2}$$

$$\theta = 60°$$

3. The critical angle for total internal refraction at a boundary between two materials is 52°. What is the Brewster's angle at this boundary?

Solution:

For the critical angle for internal refraction at a boundary between two materials with refraction indices n_1 and n_2, we have

$$\sin \theta_c = \frac{n_2}{n_1}$$

For the Brewster's angle, we have

$$\tan \theta_p = \frac{n_2}{n_1}$$

So $\tan \theta_p = \sin \theta_c$

$$\theta_p = \arctan(\sin 52°) = 38.2°$$

4. A beam of light enters a calcite prism from left as shown in Fig. 14-16. The optic axis is perpendicular to the paper. Sketch the entering and emerging beams, showing the states of polarization.

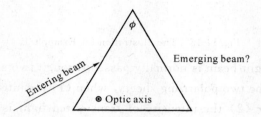

Fig. 14-16   The illustration for Example 4

Solution: See Fig. 14-17

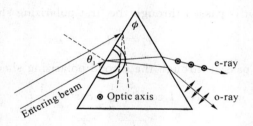

Fig. 14-17   Solution to Example 4

Questions and Problems

1. A vertically oriented, ideal polarizing sheet transmits 50% of the incident linearly polarized light. The polarizing sheet is now rotated 45°, what fraction of the incident intensity now passes? _____.

(A) 0%   (B) 50%
(C) 100%   (D) Either 0% or 100%

2. Two ideal polarizing sheets are stacked so that none of the incident unpolarized light is transmitted. A third polarizing sheet is slipped between the first two sheets at an angle of 45° to the bottom sheet. The fraction of light transmitted through the entire stack is _____.

(A) still zero   (B) 1/8   (C) 1/4   (D) 1/2

3. What would happen to the interference pattern produced from twin-slit apparatus if the light from one slit is passed through a vertically polarizing sheet and the other slit is passed through a horizontally polarizing sheet? _____.

Chapter 14　Polarization of Light

(A) The interference pattern will look like a normal double-slit pattern

(B) The interference pattern will be fainter than a normal double-slit pattern

(C) The interference pattern will be more diffuse than a normal double-slit pattern

(D) There will be no interference pattern

4. Unpolarized light strikes an air-water interface at the angle so that the reflected ray is completely polarized as shown in Fig. 14-18(a). A second ray of unpolarized light travels backwards, parallel to previously reflected ray as shown in Fig. 14-18(b). Describe the polarization of the reflected ray in Fig. 14-18(b). _____.

(A) The ray is completely polarized

(B) The ray is partial polarized

(C) The ray undergoes total internal reflection

(D) There is not enough information to answer the question

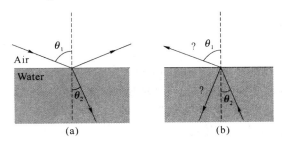

Fig. 14-18　Problem 4

5. Fig. 14-19 shows four pairs of polarizing sheets, seen face-on. Each pair is mounted in the path of initially unpolarized light. The polarizing direction of each sheet (indicated by the dashed line) is referenced to either a horizontal x axis or a vertical y axis. Rank the pairs according to the fraction of the initial intensity that they pass, greatest first. _____.

(A) (b), (d), (a), (c)　　(B) (a), (d), (b), (c)

(C) (d), (b), (a), (c)　　(D) (c), (a), (b), (d)

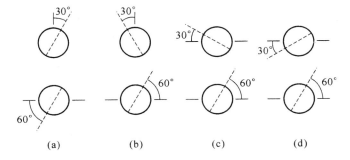

Fig. 14-19　Problem 5

6. As shown in Fig. 14-20, an initially unpolarized light is sent through three polarizing sheets whose polarizing axes make angle of $\theta_1 = 40°$, $\theta_2 = 20°$, $\theta_3 = 40°$ with the direction of the y axis. The percentage of the initial intensity which is transmitted by the system of the three sheets is _____.

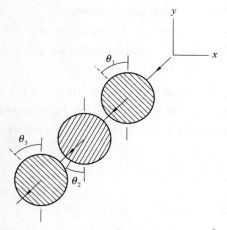

Fig. 14-20 Problem 6

7. We want to rotate the direction of polarization of a beam of plane-polarized light through 90° by sending the beam through one or more polarizing sheets. (1) What is the minimum number of sheets required? (2) What is the minimum number of sheets required if the transmitted intensity is to be more than 60% of the original intensity?

8. A beam of partially polarized light can be considered to be a mixture of polarized and unpolarized light. Suppose we send such a beam through a polarizing sheet and rotate the sheet through 360° while keeping it perpendicular to the beam. If the transmitted intensity varies by the factor of 5.0 during the rotation, what the fraction of the intensity of the original beam is associated with the polarized light?

9. When red light in vacuum is incident at the polarizing angle on a certain glass slab, the angle of refraction is 31.8°. What are (1) the index of refraction of the glass and (2) the polarizing angle?

10. The critical angle for total reflection at a boundary between two materials is 52°. What is Brewster's angle at this boundary?

11. Linearly polarized light of wavelength 525 nm strikes, at normal incidence, a wurtzite crystal (ZnS, $n_o = 2.356$, $n_e = 2.378$) that is cut with its faces parallel to the optic axis. What is the smallest possible thickness of the crystal if the emergent o-rays and e-rays combine to form linearly polarized light?

12. A narrow beam of unpolarized light falls on a calcite crystal ($n_o = 1.658$, $n_e = 1.486$) cut with its optic axis as shown in Fig. 14-21. (1) For $t = 1.12$ cm and for $\theta_i = 38.8°$, calculate the perpendicular distance between the two emerging rays x and y. (2) Which is the o-ray and which is the e-ray? (3) What are the states of polarization of the emerging rays?

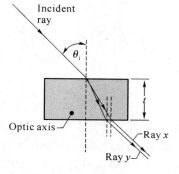

Fig. 14-21 Problem 12

Chapter 15 Special Theory of Relativity

Review of the Contents

1. Frames of references, events and transformations

Physical observers are considered to be surrounded by a **reference frame** which is a set of space and time coordinates in terms of which position or movement may be specified or with reference to which physical laws may be mathematically stated.

An **event** is a physical "happening" that occurs at a certain place and time. To record the event an observer must choose a reference frame. Different observers may use different reference frames, as shown in Fig. 15-1.

Transformations show how the description of the event in one reference frame is related to the description of the same event in another reference frame.

The theory of special relativity deals with inertial reference frames. An inertial reference frame is the one in which Newton's law of inertia is valid. Accelerating reference frames are not inertial reference frames, as shown in Fig. 15-1.

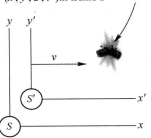

An event has space-time coordinates (x, y, z, t) in frame S and different space-time coordinates (x', y', z', t') in frame S'

Fig. 15-1 The location and time of an event are described by its space-time coordinates

2. The postulates of special relativity

The theory of special relativity is based on two postulates:

(1) First postulate (the principle of relativity): all the laws of physics are the same for observers in all inertial frames of reference.

(2) Second postulate (invariance of the speed of light): the speed of light in free space has the same value, $c(3.0 \times 10^8 \text{ m/s})$, in all inertial systems, regardless of the velocity of the observer or the velocity of the source emitting the light; that is, the speed of light is absolute.

3. The Lorentz transformations

As shown in Fig. 15-2, the Lorentz transformations relate the space and time coordinates of an event in an inertial frame S to the space and time coordinates of the same event as observed in a second inertial frame S' moving at velocity v relative to the first.

$$\begin{cases} x' = \dfrac{x - vt}{\sqrt{1 - \dfrac{v^2}{c^2}}} \\ y' = y \\ z' = z \\ t' = \dfrac{t - \dfrac{v}{c^2}x}{\sqrt{1 - \dfrac{v^2}{c^2}}} \end{cases} \qquad (15\text{-}1)$$

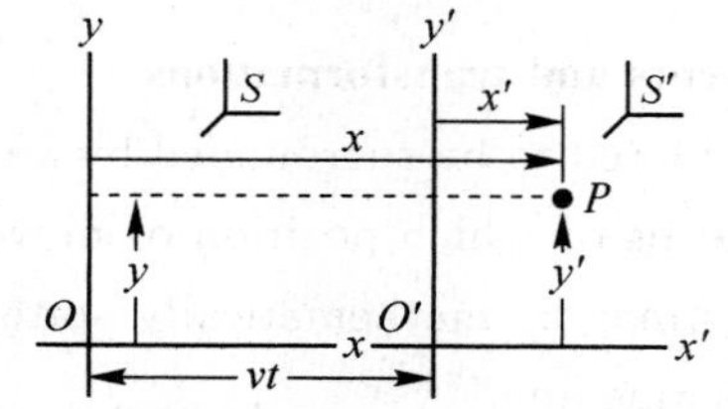

Fig. 15-2 Inertial reference frame S' moves to the right at speed v with respect to frame S

4. The relativity of time: time dilation

In a inertial reference frames, the proper time Δt_0 is the time interval between two events as measured by an observer who sees **the events occur at the same position.** Another observer who is in motion with respect to the events (i. e. in a second inertial reference frames) and who views them as occurring **at different places measures a dilated time interval** Δt. The dilated time interval is greater than the proper time interval, according to the time-dilation equation:

$$\Delta t = \dfrac{\Delta t_0}{\sqrt{1 - \dfrac{v^2}{c^2}}} \qquad (15\text{-}2)$$

In this expression, v is the relative speed between the observer who measures Δt_0 and the observer who measures Δt. Eq. 15-2 can be derived from Fig. 15-3, and also by the Lorentz transformation(Eq. 15-1).

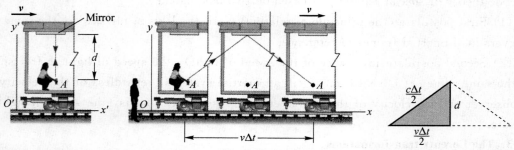

(a) A mirror is fixed to a moving vehicle, and a light pulse leaves O' at rest in the vehicle

(b) Relative to a stationary observer on Earth, the mirror and O' move with a speed v. Note that the distance the pulse travels is greater than $2d$ as measured by the stationary observer

(c) The right triangle for calculating the relationship between Δt and Δt_0

Fig. 15-3 The time dilation

Chapter 15 Special Theory of Relativity

5. The relativity of length; length contraction

The measured distance between two points depends on the frame of reference of the observer. The proper length L_0 of an object is the length of the object as measured by an observer at rest relative to the object. The length L of an object measured in a reference frame that is moving at velocity v with respect to the object is always less than the proper length L_0. This effect is known as length contraction.

$$L = L_0 \sqrt{1 - \frac{v^2}{c^2}} \qquad (15\text{-}3)$$

Eq. 15-3 can be derived by the Lorentz transformation (Eq. 15-1). The measurements of L and L_0 are shown in Fig. 15-4. Length contraction occurs only along the direction of the motion. Those dimensions that are perpendicular to the motion are not shortened.

The observer who measures the proper length may not be the observer who measures the proper time interval.

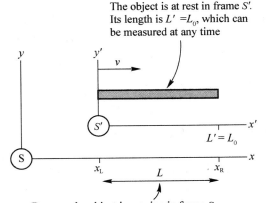

Fig. 15-4 The length contraction

6. The relativistic addition of velocities

According to special relativity, the velocity addition formula specifies how the relative velocities of moving objects are related.

$$v'_x = \frac{v_x - u}{1 - \frac{v_x u}{c^2}} \qquad v'_y = \frac{v_y}{1 - \frac{u}{c^2} v_x} \sqrt{1 - \frac{u^2}{c^2}} \qquad v'_z = \frac{v_z}{1 - \frac{u}{c^2} v_x} \sqrt{1 - \frac{u^2}{c^2}} \qquad (15\text{-}4)$$

where u is the relative velocity of the two frames of reference, $v_x (v_y, v_z)$ is the velocity of the object relative to S, and $v'_x (v'_y, v'_z)$ is the velocity of the object relative to S'. (Assume that frame S' is moving in the positive x-direction with velocity u relative to frame S.)

7. Relativistic mass and momentum

An object with a rest mass m (nonrelativistic mass), moving with speed v, has a relativistic mass m_{rel} and relativistic momentum p.

$$m_{\text{rel}} = \frac{m}{\sqrt{1 - \frac{v^2}{c^2}}}, \qquad p = m_{\text{rel}} v = \frac{mv}{\sqrt{1 - \frac{v^2}{c^2}}} \qquad (15\text{-}5)$$

Clearly, the relativistic momentum is different from the Newtonian momentum by the factor $1\big/\sqrt{1-\dfrac{V^2}{C^2}}$, which is shown in Fig. 15-5.

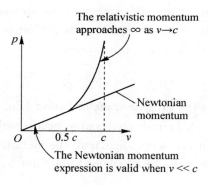

Fig. 15-5 The momentum approaches ∞ as $v \to c$

8. The equivalence of mass and energy

Energy and mass are equivalent. The total energy E of an object with a rest mass m, moving at speed v, is

$$E = m_{\text{rel}} c^2 = \frac{mc^2}{\sqrt{1-\dfrac{v^2}{c^2}}} \tag{15-6}$$

The rest energy E_0 is the total energy of an object at rest ($v=0$ m/s):

$$E_0 = mc^2 \tag{15-7}$$

An object's total energy is the sum of its rest energy and its kinetic energy K, or $E=E_0+K$. Therefore, the kinetic energy is

$$K = m_{\text{rel}} c^2 - mc^2 \tag{15-8}$$

The relativistic total energy and momentum are related according to

$$E^2 = p^2 c^2 + m^2 c^4 \tag{15-9}$$

Typical Examples

Problem solving strategy

The Lorentz coordinate transformation tells us how to relate the space-time coordinates of an event in one inertial frame of reference to the space-time coordinates of the same event in a second inertial frame. The Lorentz velocity transformation relates the velocity of an object in one inertial frame to its velocity in a second inertial frame. Execute the solution using the following steps:

(1) Determine what the target variable is.

(2) Define the two inertial frames S and S'. Remember that S' moves relative to S at a constant velocity u in the $+x$-direction.

(3) If the coordinate transformation equations are needed, make a list of space-time coordinates in the two frames, such as x_1, x'_1, t_1, t'_1, and so on. Label carefully which of

Chapter 15 Special Theory of Relativity

these you know and which you don't and use Eq. (15-1) to solve for the space-time coordinates of the event as measured in S' in terms of the corresponding values in S.

(4) In velocity-transformation problems, clearly identify the velocities u, v_x, and v'_x and use Eq. (15-4) to solve for the target variable.

Examples

1. High-energy subatomic particles coming from space interact with atoms in the earth's upper atmosphere, producing unstable particles called muons. A muon decays with a mean lifetime of 2.20×10^{-6} s as measured in a frame of reference in which it is at rest. If a muon is moving at $0.990c$ (about 2.97×10^8 m/s) relative to the earth, what will you (an observer on earth) measure its mean lifetime to be?

Solution:

This problem concerns the muon's lifetime, which is the time interval between two events: the production of the muon and its subsequent decay. This lifetime is measured by two different observers: one who observes the muon at rest and another (you) who observes it moving at $0.990c$. Let S be your frame of reference on earth, and let S' be the muon's frame of reference. The target variable is the interval between these events as measured in S. List all the space-time coordinates of the events in the two frames in Table 15-1.

Table 15-1 The space-time description in the two reference frames S and S'

S(frame of earth)	S'(frame of muon)
Event 1 (x_1, t_1)	(x'_1, t'_1)
Event 2 (x_2, t_2)	(x'_2, t'_2)
$x_1 \neq x_2$	$x'_1 = x'_2$
$\Delta t = t_2 - t_1 = ?$	$\Delta t_0 = t'_2 - t'_1 =$ (proper time) $= 2.20 \times 10^{-6}$ s

From Eq. (15-2)

$$\Delta t = \frac{\Delta t_0}{\sqrt{1 - \frac{v^2}{c^2}}} = \frac{2.20 \times 10^{-6} \text{ s}}{\sqrt{1 - (0.990)^2}} = 15.6 \times 10^{-6} \text{ s}$$

The result predicts that the mean lifetime of the muon in the earth frame (Δt) is about seven times longer than in the muon's frame (Δt_0). This prediction has been verified experimentally.

2. As was stated in Example 1, a muon has, on average, a proper lifetime of 2.20×10^{-6} s and a dilated lifetime of 15.6×10^{-6} s in a frame in which its speed is $0.990c$. Multiplying constant speed by time to find distance gives $0.990(3.00 \times 10^8 \text{ m/s}) \times (2.20 \times 10^{-6} \text{ s}) = 653$ m and $0.990(3.00 \times 10^8 \text{ m/s}) \times (15.6 \times 10^{-6} \text{ s}) = 4,630$ m. Interpret these two distances.

Solution:

If an average muon moves at $0.990c$ past observers, they will measure it to be created at one point and then to decay 15.6×10^{-6} s later at another point 4,630 m away. For example, this muon could be created level with the top of a mountain and then move straight down to decay at its base 4,630 m below. However, an observer moving with an average muon will say that it traveled only 653 m because it existed for only 2.20×10^{-6} s. To show that this answer is completely consistent, consider the mountain. The 4,630 m distance is its height, a proper length in the direction of motion. Relative to the observer traveling with this muon, the mountain moves up at $0.990c$ with the 4,630 m length contracted to

$$l = l_0 \sqrt{1 - \frac{v^2}{c^2}} = 4{,}630 \text{ m} \times \sqrt{1 - 0.990^2} = 653 \text{ m}$$

Thus we see that length contraction is consistent with time dilation.

3. Relative velocities. (1) As shown in Fig. 15-6, a spaceship moving away from the earth with speed $0.900c$ fires a robot space probe in the same direction as its motion, with speed $0.700c$ relative to the spaceship. What is the probe's velocity relative to the earth? (2) A scout ship tries to catch up with the spaceship by traveling at $0.950c$ relative to the earth. What is the velocity of the scout ship relative to the spaceship?

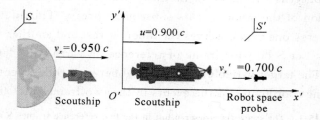

Fig. 15-6 The illustration for Example 3

Solution:

This example uses the Lorentz velocity transformation. Let the earth's frame of reference be S, and let the spaceship's frame of reference be S'. The relative velocity of the two frames is $u = 0.900c$. The target variable in part (1) is the velocity of the probe relative to S; the target variable in part (2) is the velocity of the scout ship relative to S'.

(1) We are given the velocity of the probe relative to the spaceship (in S'), $v'_x = 0.700c$. We use Eq. (15-4) to determine its velocity v_x, relative to the earth (in S):

$$v_x = \frac{v'_x + u}{1 + \frac{v'_x u}{c^2}} = \frac{0.700c + 0.900c}{1 + \frac{(0.700c) \times (0.900c)}{c^2}} = 0.982c$$

(2) We are given the velocity of the scout ship relative to the earth (in S), $v_x = 0.950c$. We determine its velocity v'_x relative to the spaceship (in S):

$$v'_x = \frac{v_x - u}{1 - \frac{v_x u}{c^2}} = \frac{0.950c - 0.900c}{1 - \frac{(0.950c) \times (0.900c)}{c^2}} = 0.345c$$

It is instructive to compare our results to what we would have obtained if we had used the Galilean velocity transformation formula. In part (1) we would have found the probe's velocity relative to the earth to be $v_x = v'_x + u = 0.700c + 0.900c = 1.600c$. This value is greater than the speed of light and so must be incorrect. In part (2) we would have found the scout ship's velocity relative to the spaceship to be $v'_x = v_x - u = 0.950c - 0.900c = 0.050c$; the relativistically correct value, $v'_x = 0.345c$, is almost seven times greater than the incorrect Galilean value.

Chapter 15 Special Theory of Relativity

4. The fission, or splitting, of uranium was discovered in 1938 by Lise Meitner, who successfully interpreted some curious experimental results found by Otto Hahn as due to fission. (Hahn received the Nobel prize) The fission of $^{235}_{92}U$ begins with the absorption of a slow-moving neutron that produces an unstable nucleus of ^{236}U. The ^{236}U nucleus then quickly decays into two heavy fragments moving at high speed, as well as several neutrons. Most of the kinetic energy released in such a fission is carried off by the two large fragments. (1) For the typical fission process

$$^1_0n + {}^{235}_{92}U \rightarrow {}^{141}_{56}Ba + {}^{92}_{36}Kr + 3{}^1_0n$$

calculate the kinetic energy in MeV carried off by the fission fragments. (2) What percentage of the initial energy is converted into kinetic energy? The atomic masses involved are given below in atomic mass units.

$^1_0n = 1.008,665u$ $^{235}_{92}U = 235.043,924u$ $^{141}_{56}Ba = 140.903,496u$ $^{92}_{36}Kr = 91.907,936u$

Solution:

The goal of the problem is to understand the production of energy from nuclear sources. Write the conservation law as a sum of kinetic energy and rest energy, solve for the final kinetic energy and v, and then yields the speeds.

(1) Calculate the final kinetic energy for the given process.

Apply the conservation of relativistic energy equation,

$$(K + mc^2)_{initial} = (K + mc^2)_{final}$$

Assuming that $K_{initial} = 0$,

$$0 + m_n c^2 + m_U c^2 = m_{Ba} c^2 + m_{Kr} c^2 + 3m_n c^2 + K_{final}$$

$$K_{final} = [(m_n + m_U) - (m_{Ba} + m_{Kr} + 3m_n)]c^2$$

$$= 0.215,162\ uc^2$$

$$= 0.215,162 \times 931.5\ \text{MeV}$$

$$= 200.422\ \text{MeV}$$

(2) What percentage of the initial energy is converted into kinetic energy?

Compute the total energy, which is the initial energy:

$$E_{initial} = 0 + m_n c^2 + m_U c^2 = 2.198,82 \times 10^5\ \text{MeV}$$

Divide the kinetic energy by the total energy and multiply by 100%:

$$\frac{200.422\ \text{MeV}}{2.198,82 \times 10^5\ \text{MeV}} \times 100\% = 9.115 \times 10^{-2}\%$$

Remarks: The calculation shows that in the nuclear reaction only about one thousandth of the rest energy of the constituent particles is released. The energy released in some nuclear fusions is several times as many as the fission energy.

Questions and Problems

1. Two events occur simultaneously on the x axis of reference frame S, one at $x = -a$ and the other at $x = +a$. According to an observer moving in the positive x direction _____.

(A) the event at $x = +a$ occurs first

(B) the event at $x=-a$ occurs first

(C) either event might occur first, depending on the value of a and the observer's speed

(D) the events are simultaneous

2. Spaceship A, traveling past us at $0.7c$, sends a message capsule to spaceship B, which is in front of A and is traveling in the same direction as A at $0.8c$ relative to us. The capsule travels at $0.95c$ relative to us. A clock that measures the proper time between the sending and receiving of the capsule travels _____.

(A) in the same direction as the spaceships at $0.7c$ relative to us

(B) in the opposite direction from the spaceships at $0.7c$ relative to us

(C) in the same direction as the spaceships at $0.8c$ relative to us

(D) in the same direction as the spaceships at $0.95c$ relative to us

3. If the proper distance to a star is 5.0 light-years, what is the distance as measured in a spaceship going to the star at $0.8c$? _____.

(A) 6.3 light-years   (B) 5.0 light-years

(C) 4.0 light-years   (D) 3.0 light-years

4. With respect to the earth, object 1 is moving at speed $0.90c$ to the right. Object 2 is moving in the same direction at speed $0.70c$ with respect to object 1. How fast is object 2 moving with respect to the earth? _____.

(A) $0.64c$ (B) $0.80c$ (C) $0.98c$ (D) $1.6c$

5. The work that must be done to increase the speed of an electron ($m=9.11 \times 10^{-31}$ kg) from $0.90c$ to $0.95c$ is _____.

(A) 2.6×10^{-13} J   (B) 8.2×10^{-13} J

(C) 3.2×10^{-13} J   (D) 7.4×10^{-14} J

6. One proton is moving at speed $0.4c$ and another proton is moving at speed $0.8c$. The ratio of the momentum of the faster moving proton to that of the slower is _____.

(A) 2.0   (B) 3.1   (C) 4.0   (D) 4.7

7. The positive muon, an unstable particle, lives on average 2.20×10^{-6} s (measured in its own frame of reference) before decaying. If such a particle is moving, with respect to the laboratory, with a speed of $0.900c$, the average lifetime measured in the laboratory is _____. The average distance that the particle moves before decaying, measured in the laboratory, is _____.

8. A friend passes by you in a spacecraft traveling at a high speed. He tells you that his craft is 20.0 m long and that the identically constructed craft you are sitting in is 19.0 m long. According to your observations, the length of your spacecraft is _____, the length of your friend's craft is _____, and the speed of your friend's craft is _____.

9. An observer in frame S' is moving to the right ($+x$-direction) at speed $u=0.600c$ away from a stationary observer in frame S. The observer in S' measures the speed v' of a particle moving to the right away from her. If $v'=0.400c$, the speed v that the observer in S measures for the particle is _____.

Chapter 15 Special Theory of Relativity

10. An unstable particle at rest breaks into two fragments of unequal mass. The mass of the first fragment is 2.50×10^{-28} kg, and that of the other is 1.67×10^{-27} kg. If the lighter fragment has a speed of $0.893c$ after the breakup, the speed of the heavier fragment is _____.

11. The power output of the Sun is 3.77×10^{26} W. The mass converted to energy in the Sun each second is _____.

12. A proton (rest mass 1.67×10^{-27} kg) has total energy that is 4.00 times its rest energy. The kinetic energy of the proton is _____. The magnitude of the momentum of the proton is _____. The speed of the proton is _____.

13. A moving rod is observed to have a length of 2.00 m and to be oriented at an angle of 30.0° with respect to the direction of motion, as shown in Fig. 15-7. The rod has a speed of $0.995c$. (1) What is the proper length of the rod? (2) What is the orientation angle in the proper frame?

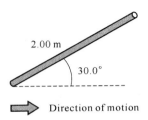

Fig. 15-7 The illustration for Problem 13

14. A spacecraft is launched from the surface of the earth with a velocity of $0.600c$ at an angle of 50.0° above the horizontal positive x axis. Another spacecraft is moving past, with a velocity of $0.700c$ in the negative x direction. Determine the magnitude and direction of the velocity of the first spacecraft as measured by the pilot of the second spacecraft.

15. A pion at rest ($m_\pi = 273 m_e$) decays to a muon ($m_\mu = 207 m_e$) and an antineutrino ($m_{\bar{\nu}} \approx 0$). The reaction is written $\pi \rightarrow \mu + \bar{\nu}$. Find the kinetic energy of the muon and the energy of the antineutrino in electron volts.

Chapter 16 Fundamentals of Quantum Theory

Review of the Contents

1. Blackbody radiation and Planck's hypothesis

The characteristics of **blackbody radiation** can't be explained with classical concepts. As shown in Fig. 16-1, the peak of a blackbody radiation curve is given by **Wien's displacement law**,

$$\lambda_{max} T = 0.289,8 \times 10^{-2} \text{ m} \cdot \text{K} \tag{16-1}$$

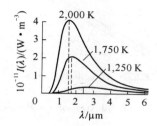

Fig. 16-1 Blackbody radiation

where λ_{max} is the wavelength at which the curve peaks and T is the absolute temperature of the object emitting the radiation.

Planck first introduced the quantum concept when he assumed that the subatomic oscillators responsible for blackbody radiation could have only discrete amounts of energy given by

$$E_n = nhf \tag{16-2}$$

where n is a positive integer called a **quantum number** and f is the frequency of vibration of the resonator.

2. The photoelectric effect and the particle theory of light

As shown in Fig. 16-2, the **photoelectric effect** is a process whereby electrons are ejected from a metal surface when light is incident on that surface. Einstein provided a successful explanation of this effect by extending Planck's quantum hypothesis to electromagnetic waves. In this model, light is viewed as a stream of particles called **photons**, each with energy $E = hf$, where f is the light frequency and h is **Planck's constant**. The maximum kinetic energy of the ejected photoelectrons is

$$K_{max} = hf - \phi \tag{16-3}$$

where ϕ is the **work function** of the metal.

Chapter 16   Fundamentals of Quantum Theory

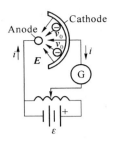

Fig. 16-2   Photoelectric effect

3. The Compton effect

As shown in Fig. 16-3, X-rays from an incident beam are scattered at various angles by electrons in a target such as carbon. In such a scattering event, a shift in wavelength is observed for the scattered X-rays. This phenomenon is known as the **Compton shift**. Conservation of momentum and energy applied to a photon-electron collision yields the following expression for the shift in wavelength of the scattered X-rays:

$$\Delta\lambda = \lambda - \lambda_0 = \frac{h}{m_e c}(1 - \cos\theta) \qquad (16\text{-}4)$$

Here, m_e is the mass of the electron, c is the speed of light, and θ is the scattering angle.

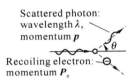

Fig. 16-3   Compton effect

4. The dual nature of light and matter

Light exhibits both a particle and a wave nature. De Broglie proposed that all matter has both a particle and a wave nature. The **de Broglie wavelength** of any particle of mass m and speed v is

$$\lambda = \frac{h}{p} = \frac{h}{mv} \qquad (16\text{-}5)$$

De Broglie also proposed that the frequencies of the waves associated with particles obey the Einstein relationship

$$E = hf \qquad (16\text{-}6)$$

5. The wave function

In the theory of quantum mechanics, each particle is described by a quantity Ψ called the **wave function**. The probability per unit volume of finding the particle at a particular point at some instant is proportional to $|\Psi|^2$. Quantum mechanics has been highly successful in describing the behavior of atomic and molecular systems.

6. The uncertainty principle

According to Heisenberg's **uncertainty principle**, it is impossible to measure simultaneously the exact position and exact momentum of a particle. As shown in Fig. 16-4, if Δx is the uncertainty in the measured position and Δp_x the uncertainty in the momentum, the product $\Delta x \Delta p_x$ is given by

$$\Delta x \Delta p_x \geqslant \frac{\hbar}{2}\frac{h}{4\pi} \quad \text{or} \quad \Delta x \Delta p_x \geqslant \hbar \qquad (16\text{-}7)$$

where $\hbar = h/2\pi$.

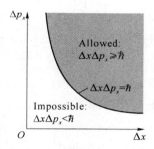

Fig. 16-4   Uncertainty principle

7. The Bohr theory of hydrogen and modification of the Bohr theory

The Bohr model of the atom is successful in describing the spectra of atomic hydrogen and hydrogen-like ions. One of the basic assumptions of the model is that the electron can exist only in certain orbits such that its angular momentum mvr is an integral multiple of $\hbar$, where $\hbar$ is Planck's constant divided by 2π. Assuming circular orbits and a Coulomb force of attraction between electron and proton, the energies of the quantum states for hydrogen are

$$E_n = -\frac{m_e k_e^2 e^4}{2\hbar^2}\left(\frac{1}{n^2}\right) \quad n=1,2,3,\cdots \qquad (16\text{-}8)$$

where k_e is the Coulomb constant, e is the charge on the electron, and n is an integer called a **quantum number**.

If the electron in the hydrogen atom jumps from an orbit having quantum number n_i to an orbit having quantum number n_f, it emits a photon of frequency f, given by

$$f = \frac{m_e k_e^2 e^4}{4\pi \hbar^3}\left(\frac{1}{n_f^2}-\frac{1}{n_i^2}\right) \qquad (16\text{-}9)$$

Bohr's correspondence principle states that quantum mechanics is in agreement with classical physics when the quantum numbers for a system are very large.

The Bohr theory can be generalized to hydrogen-like atoms, such as singly ionized helium or doubly ionized lithium. This modification consists of replacing e^2 by Ze^2 wherever it occurs.

8. Quantum mechanics and the hydrogen atom & the spin magnetic quantum number

One of the many successes of quantum mechanics is that the quantum numbers n, l, and m_l associated with atomic structure arise directly from the mathematics of the theory. The quantum number n is called the **principal quantum number**, l is the **orbital quantum number**, and m_l is the **orbital magnetic quantum number**. As shown in Fig. 16-5, these quantum numbers can take only certain values: $1 \leqslant n < \infty$ in integer steps, $0 \leqslant l \leqslant n-1$, and $-l \leqslant m_l \leqslant l$.

Chapter 16 Fundamentals of Quantum Theory

In addition, the fourth quantum number, called the **spin magnetic quantum number** m_s, is needed to explain a fine doubling of lines in atomic spectra(Fig. 16-6), with $m_s = \pm \frac{1}{2}$. An electron has not only the orbital angular momentum, indicated by l and m_l, but also the spin angular momentum, indicated by $s = \frac{1}{2}$ and $m_s = \pm \frac{1}{2}$ and spin down when $m_s = \frac{1}{2}$. When an atom is placed in a magnetic field B, one energy level E_s of the atom will split into two energy states because of the electron spin as shown in Fig. 16-6.

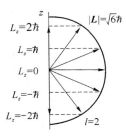
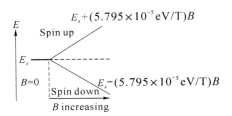

Fig. 16-5 Orbital magnetic quantum number

Fig. 16-6 Electron spin

9. The exclusion principle and the periodic table

An understanding of the periodic table of the elements became possible when Pauli formulated the **exclusion principle**, which states that no two electrons in the same atom can have the same values for the set of quantum numbers n, l, m_l, and m_s. A particular set of these quantum numbers is called a quantum state. The exclusion principle explains how different energy levels in atoms are populated. Once one subshell is filled, the next electron goes into the vacant subshell that is lowest in energy. Atoms with similar configurations in their outermost shell have similar chemical properties and are found in the same column of the periodic table.

10. Atomic transitions

When an atom is irradiated by light of all different wavelengths, it will only absorb wavelengths equal to the difference in energy of two of its energy levels, as shown in Fig. 16-7(a). This phenomenon, called **stimulated absorption**, places an atom's electrons into excited states. Atoms in an excited state have a probability of returning to a lower level of excitation by **spontaneous emission**, as shown in Fig. 16-7(b). The wavelengths that can be emitted are the same as the wavelengths that can be absorbed. If an atom is in an excited state and a photon with energy $hf = E_2 - E_1$ is incident on it, the probability of emission of a second photon of this energy is greatly enhanced. The emitted photon is exactly in phase with the incident photon. This process is called **stimulated emission**, as shown in Fig. 16-7(c). The emitted and original photon can then stimulate more emission, creating an amplifying effect.

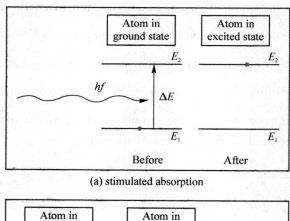
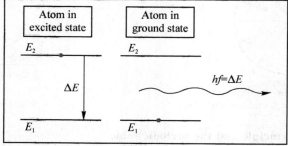
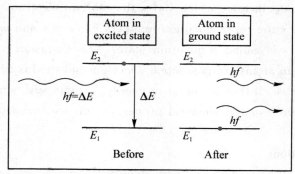

Fig. 16-7　Atomic transitions

Typical Examples

Problem solving strategy

1. For an individual photon, its energy is proportional to the frequency(Eq. 16-6), and its momentum is inversely proportional to the wavelength(Eq. 16-5). These concepts, along with the idea of stopping potential in the photoelectric effect, allow you to solve almost any problem involving photons. Decide what your target variable is. It could be the photon's wavelength λ, frequency f, energy E, or momentum p. If the problem involves the

Chapter 16 Fundamentals of Quantum Theory

photoelectric effect, the target variable could be the maximum kinetic energy of photoelectrons K_{max}, the stopping potential V_0, or the work function ϕ.

2. Many problems in atomic structure can be solved just by reference to the quantum numbers that describe the total energy E, the magnitude of the orbital angular momentum L, the z-component of L, and other properties of an atom. Decide what the target variable is and collect the equations that you will use to determine its value. Be sure you know the possible values of the quantum numbers n, l, and m_l, for the hydrogen atom. They are all integers; n is always greater than zero, l can be zero or positive up to $(n-1)$, and m_l, can range from $-l$ to l. Be able to count the number of (n, l, m_l) states in each shell (K, L, M, and so on) and subshell ($3s$, $3p$, $3d$, and so on). When you check numerical values, it's helpful to be familiar with some typical magnitudes. In addition, the quantum number m_s describes the electron spin, with spin up when $m_s = +\frac{1}{2}$ and spin down when $m_s = -\frac{1}{2}$.

Examples

1. A sodium surface is illuminated with light of wavelength 0.300 μm. The work function for sodium is 2.46 eV. (1) Calculate the energy of each photon in electron volts, (2) the maximum kinetic energy of the ejected photoelectrons, and (3) the cutoff wavelength for sodium.

Solution:

(1) Calculate the energy of each photon.

Obtain the frequency from the wavelength:

$$c = f\lambda \Rightarrow f = \frac{c}{\lambda} = \frac{3.00 \times 10^8 \text{ m/s}}{0.300 \times 10^{-6} \text{ m}} = 1.00 \times 10^{15} \text{ Hz}$$

Calculate the photon's energy:

$$E = hf = (6.63 \times 10^{-34} \text{ J} \cdot \text{s})(1.00 \times 10^{15} \text{ Hz}) = 6.63 \times 10^{-19} \text{ J}$$

$$= (6.63 \times 10^{-19} \text{ J}) \left(\frac{1.00 \text{ eV}}{1.60 \times 10^{-19} \text{ J}} \right) = 4.14 \text{ eV}$$

(2) Find the maximum kinetic energy of the photoelectrons.

$$K_{max} = hf - \phi = 4.14 \text{ eV} - 2.46 \text{ eV} = 1.68 \text{ eV}$$

(3) Compute the cutoff wavelength.

Convert ϕ from electron volts to joules:

$$\phi = 2.46 \text{ eV} = (2.46 \text{ eV})(1.60 \times 10^{-19} \text{ J/eV}) = 3.94 \times 10^{-19} \text{ J}$$

Find the cutoff wavelength

$$\lambda_c = \frac{hc}{\phi} = \frac{(6.63 \times 10^{-34} \text{ J} \cdot \text{s})(3.00 \times 10^8 \text{ m/s})}{3.94 \times 10^{-19} \text{ J}} = 5.05 \times 10^{-7} \text{ m} = 505 \text{ nm}$$

The cutoff wavelength is in the yellow-green region of the visible spectrum.

2. For a certain cathode material in a photoelectric-effect experiment, you measure a stopping potential of 1.0 V for light of wavelength 600 nm, 2.0 V for 400 nm, and 3.0 V for 300 nm. Determine the work function for this material and the value of Planck's constant.

Solution:

$$V_0 = \frac{h}{e}f - \frac{\phi}{e}$$

In this form we see that the slope of the line is h/e and the intercept on the vertical axis (corresponding to $f=0$) is at $-\phi/e$. The frequencies, obtained from $f = c/\lambda$ and $c = 3.00 \times 10^8$ m/s, are 0.50×10^{15} Hz, 0.75×10^{15} Hz, and 1.0×10^{15} Hz, respectively. The graph is shown in Fig. 16-8. From it we find

$$-\frac{\phi}{e} = \text{vertical intercept} = -1.0 \text{ V}$$

$$\phi = 1.0 \text{ eV} = 1.6 \times 10^{-19} \text{ J}$$

and

$$\text{slope} = \frac{\Delta V_0}{\Delta f} = \frac{3.0 \text{ V} - (-1.0 \text{ V})}{1.00 \times 10^{15} \text{ s}^{-1} - 0} = 4.0 \times 10^{-15} \text{ J} \cdot \text{s/C}$$

$$h = \text{slope} \times e = (4.0 \times 10^{-15} \text{ J} \cdot \text{s/C})(1.6 \times 10^{-19} \text{ C}) = 6.4 \times 10^{-34} \text{ J} \cdot \text{s}$$

This experimental value differs by about 3% from the accepted value.

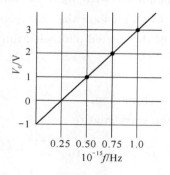

Fig. 16-8 The illustration for Example 2

3. Consider the wave function $\psi(x) = A_1 e^{ikx} + A_2 e^{-ikx}$, where k is positive. Is this a valid stationary-state wave function for a free particle? What is the energy?

Solution:

Substituting $\psi(x) = A_1 e^{ikx} + A_2 e^{-ikx}$ and $U(x) = 0$ into the Schrödinger equation, we obtain

$$-\frac{\hbar^2}{2m} \frac{d^2 \psi(x)}{dx^2} = -\frac{\hbar^2}{2m} \frac{d^2 (A_1 e^{ikx} + A_2 e^{-ikx})}{dx^2}$$

$$= -\frac{\hbar^2}{2m} [(ik)^2 A_1 e^{ikx} + (-ik)^2 A_2 e^{-ikx}]$$

$$= \frac{\hbar^2 k^2}{2m} (A_1 e^{ikx} + A_2 e^{-ikx})$$

$$= \frac{\hbar^2 k^2}{2m} \psi(x)$$

We see that the right-hand side of the Schrödinger equation is equal to the product of a

Chapter 16　Fundamentals of Quantum Theory

constant and $\psi(x)$, so this $\psi(x)$ is indeed a valid stationary-state wave function for a free particle. The constant is just equal to the energy: $\hbar^2 k^2/2m$.

4. A particle is described by the wave function

$$\psi(x) = \begin{cases} 0 & x \leqslant 0 \\ ce^{-x/L} & x \geqslant 0 \end{cases}$$

where $L = 1$ nm. (1) Determine the value of the constant c. (2) Determine the probability density $P(x)$. (3) Calculate the probability of finding the particle in the region $x \geqslant 1$ nm.

Solution:

(1) The wave function is an exponential $\psi(x) = ce^{-x/L}$ that extends from $x = 0$ to $x = +\infty$. The normalization condition is

$$1 = \int_{-\infty}^{\infty} |\psi(x)|^2 dx = c^2 \int_0^{\infty} e^{-2x/L} dx = -\frac{c^2 L}{2} e^{-2x/L} \Big|_0^{\infty} = \frac{c^2}{2L}$$

We can solve this for the normalization constant c:

$$c = \sqrt{\frac{2}{L}} = \sqrt{\frac{2}{1 \text{ nm}}} = 1.414 \text{ nm}^{-1/2}$$

(2) The probability density is

$$P(x) = |\psi(x)|^2 = (2.0 \text{ nm}^{-1}) e^{-2x/(1.0 \text{ nm})}$$

(3) The probability of finding the particle in the region $x \geqslant 1$ nm is the shaded area under the probability density curve. We must integrate to find a numerical value. The probability is

$$\text{Prob}(x \geqslant 1 \text{ nm}) = \int_{1 \text{ nm}}^{\infty} |\psi(x)|^2 dx = (2.0 \text{ nm}^{-1}) \int_{1 \text{ nm}}^{\infty} e^{-2x/(1.0 \text{ nm})} dx = e^{-2} = 13.5\%$$

There is a 13.5% chance of finding the particle beyond 1 nm and thus an 86.5% chance of finding it within the interval $0 \leqslant x \leqslant 1$ nm as shown in Fig. 16-9. Unlike classical physics, we cannot make an exact prediction of the particle's position.

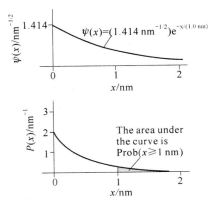

Fig. 16-9　The wave function and the probability density in Example 4

Questions and Problems

1. A photon in light beam A has twice the energy of a photon in light beam B. The ratio p_A/p_B of their momenta is _____.

    (A) 1/2   (B) 1/4   (C) 1   (D) 2

2. The stopping potential for electrons ejected by 6.8×10^{14} Hz electromagnetic radiation incident on a certain sample is 1.8 V. The kinetic energy of the most energetic electrons ejected and the work function of the sample, respectively, are _____.

    (A) 1.8 eV, 2.8 eV   (B) 1.8 eV, 1.0 eV
    (C) 1.8 eV, 4.6 eV   (D) 2.8 eV, 1.0 eV

3. Electromagnetic radiation with a wavelength of 3.5×10^{-12} m is scattered from stationary electrons and photons that have been scattered through 50° are detected. After a scattering event the magnitude of the electron's momentum is _____.

    (A) 0   (B) 1.5×10^{-22} kg·m/s
    (C) 2.0×10^{-22} kg·m/s   (D) 2.2×10^{-22} kg·m/s

4. Consider the following three particles
   (1) a free electron with kinetic energy K_0
   (2) a free proton with kinetic energy K_0
   (3) a free proton with kinetic energy $2K_0$
   Rank them according to the wavelengths of their matter waves, from least to greatest. _____.

    (A) 1, 2, 3   (B) 3, 2, 1   (C) 2, 3, 1   (D) 1, 3, 2

5. $\psi(x)$ is the wave function for a particle moving along the x axis. The probability that the particle is in the interval from $x=a$ to $x=b$ is given by _____.

    (A) $|\psi(b)|/|\psi(a)|$ (B) $|\psi(b)|^2/|\psi(a)|^2$
    (C) $\int_a^b \psi(x)\,dx$ (D) $\int_a^b |\psi(x)|^2\,dx$

6. The uncertainty in position of an electron in a certain state is 5×10^{-10} m. The uncertainty in its momentum might be _____.

    (A) 5.0×10^{-24} kg·m/s   (B) 4.0×10^{-24} kg·m/s
    (C) 3.0×10^{-24} kg·m/s   (D) all of the above

7. An electron in an atom initially has an energy 7.5 eV above the ground state energy. It drops to a state with an energy of 3.2 eV above the ground state energy and emits a photon in the process. The momentum of the photon is _____.

    (A) 1.7×10^{-27} kg·m/s   (B) 2.3×10^{-27} kg·m/s
    (C) 4.0×10^{-27} kg·m/s   (D) 5.7×10^{-27} kg·m/s

8. Which of the following sets of quantum numbers is possible for an electron in a hydrogen atom? _____.

    (A) $n=4, l=3, m_l=-3$ (B) $n=4, l=4, m_l=-2$

(C) $n=5$, $l=-1$, $m_l=2$ (D) $n=3$, $l=1$, $m_l=-2$

9. An atom is in a state with orbital quantum number $l=2$. Possible values of the magnetic quantum number m_l are _____.

(A) 0, 1, 2   (B) 0, 1
(C) -1, 0, 1   (D) -2, -1, 0, 1, 2

10. An electron is in a quantum state for which there are seven allowed values of the z component of the angular momentum. The magnitude of the angular momentum is _____.

(A) $\sqrt{3}\hbar$ (B) $\sqrt{7}\hbar$ (C) $\sqrt{9}\hbar$ (D) $\sqrt{12}\hbar$

11. When ultraviolet light with a wavelength of 254 nm falls on a clean copper surface, the stopping potential necessary to stop emission of photoelectrons is 0.181 V. The photoelectric threshold wavelength for this copper surface is _____ and the work function for this surface is _____.

12. A hydrogen atom initially in the ground level absorbs a photon, which excites it to the $n=4$ level. The wave-length and frequency of the photon is _____ and _____, respectively.

13. Two stars, both of which behave like ideal blackbodies, radiate the same total energy per second. The cooler one has a surface temperature T and 3.0 times the diameter of the hotter star. The temperature of the hotter star in terms of T is _____. The ratio of the peak-intensity wavelength of the hot star to the peak-intensity wavelength of the cool star is _____.

14. The x-coordinate of an electron is measured with an uncertainty of 0.20 nm. The x-component of the electron's velocity, v_x, is _____, if the minimum percentage uncertainty in a simultaneous measurement of v_x is 1.0%.

15. Consider a wave function given by $\psi(x)=A\sin kx$, where $k=2\pi/\lambda$ and A is a real constant. (1) For $x=$ _____, there is the highest probability of finding the particle described by this wave function. (2) For $x=$ _____, the probability is zero.

16. X-rays are produced in a tube operating at 18.0 kV. After emerging from the tube, X-rays with the minimum wavelength produced strike a target and are Compton-scattered through an angle of 45.0°. (1) What is the original X-ray wavelength? (2) What is the wavelength of the scattered X-rays? (3) What is the energy of the scattered X-rays (in electron volts)?

17. (1) What is the smallest amount of energy in electron volts that must be given to a hydrogen atom initially in its ground level so that it can emit the H_α line in the Balmer series? (2) How many different possibilities of spectral-line emissions are there for this atom when the electron starts in the $n=3$ level and eventually ends up in the ground level? Calculate the wavelength of the emitted photon in each case.

18. What must be the temperature of an ideal blackbody so that photons of its radiated light having the peak-intensity wavelength can excite the electron in the Bohr-model hydrogen atom from the ground state to the third excited state?

19. A beam of protons and a beam of alpha particles (of mass 6.64×10^{-27} kg and charge

+2e) are accelerated from rest through the same potential difference and pass through identical circular holes in a very thin, opaque film. When viewed far from the hole, the diffracted proton beam forms its first dark ring at 15° with respect to its original direction. When viewed similarly, at what angle will the alpha particle form its first dark ring?

20. A particle is described by the normalized wave function $\psi(x,y,z) = A e^{-\alpha(x^2+y^2+z^2)}$, where A and α are real, positive constants. (1) Determine the probability of finding the particle at a distance between r and $r+dr$ from the origin. (2) For what value of r does the probability in part (1) have its maximum value? Is this the same value of r for which $|\psi(x,y,z)|^2$ is a maximum? Explain any difference.

Answers to Questions and Problems

Chapter 1

1. (1) (D)   (2) (B)   2. (C)   3. (1) (C)   (2) (A)   4. (C)
5. (D)   6. (C)   7. (B)   8. (B)   9. (D)   10. (D)

11. **Answer:** $v = 4\hat{i}$ m/s; $r = 2.67\hat{i}$ m

12. **Answer:** $v = 2\sqrt{x+x^3}$

13. **Answer:** $v = -18.0\sin 3.0t\,\hat{i} + 18.0\cos 3.0t\,\hat{j}$; $a = -(54\cos 3.0t\,\hat{i} + 54\sin 3.0t\,\hat{j})$; $x^2 + y^2 = 36$; $a = -9r$

14. **Answer:** $a_t = \dfrac{dv}{dt} = \dfrac{g^2 t}{\sqrt{v_0^2 + (gt)^2}}\Big|_{t=5} = 8.58$ m/s^2; $a_n = \sqrt{a^2 - a_t^2} \approx 5.14$ m/s^2

15. **Answer:**

(1) the magnitude of the velocity $v = \dfrac{dy}{dt} = bt$, direction is along tangent;

(2) $a_t = \dfrac{dv}{dt}\hat{\tau} = b\hat{\tau}$, $a_n = \dfrac{v^2}{R}\hat{n} = \dfrac{b^2 t^2}{R}\hat{n}$, so $a = \dfrac{b^2 t^2}{R}\hat{n} + b\hat{\tau}$

16. **Answer:** $v = 7.5\hat{j}$ m/s; $r = (4.17\hat{i} + 6.25\hat{j})$ m

17. **Answer:**

(1) Its path equation $x^2 + y^2 = R^2$, So it moves in a circle of radius R;

(2) $v_x = \dfrac{dx}{dt} = -\omega R \sin \omega t$, $v_y = \dfrac{dy}{dt} = \omega R \cos \omega t$, $v = \sqrt{v_x^2 + v_y^2} = \omega R$;

(3) $a = \dfrac{dv_x}{dt}\hat{i} + \dfrac{dv_y}{dt}\hat{j} = -\omega^2 R\cos\omega t\,\hat{i} - \omega^2 R\sin\omega t\,\hat{j} = -\omega^2 r$ is opposite to the position vector $a = \sqrt{a_x^2 + a_y^2} = \omega^2 R$

18. **Answer:**

At point B, g is the normal acceleration. Therefore $\rho_B = \dfrac{v_x^2}{g} = \dfrac{v_0^2 \cos^2 \alpha}{g}$;

At point C, $a_t = g\sin\theta$, $a_n = g\cos\theta$, $v = \dfrac{v_x}{\cos\theta} = \dfrac{v_0\cos\alpha}{\cos\theta}$, $\rho_C = \dfrac{v^2}{a_n} = \dfrac{v_x^2}{g\cos^3\theta} = \dfrac{v_0^2 \cos^2\alpha}{g\cos^3\theta}$

19. **Answer:**

(1) Motional equation $r = \dfrac{1}{2}bv_0 t^2\,\hat{i} + v_0 t\,\hat{j}$;

(2) The path equation of balloon $x = \dfrac{b}{2v_0}y^2$;

(3) The tangential acceleration $a_t = \dfrac{dv}{dt} = \dfrac{b^2 v_0 t}{\sqrt{b^2 t^2 + 1}} = \dfrac{b^2 v_0 y}{\sqrt{b^2 y^2 + v_0^2}}$, $\rho = \dfrac{v^2}{a_n} = \dfrac{(b^2 y^2 + v_0^2)^{3/2}}{bv_0^2}$

20. **Answer**: $v = \dfrac{mg}{b}(1-e^{-\frac{b}{m}t})$

21. **Answer**: BG means boat to ground, BW means boat to water, WG means water to ground. (1) $\boldsymbol{v}_{BG} = \boldsymbol{v}_{BW} + \boldsymbol{v}_{WG}$, $v_{BG_x} = v_{BW_x} + v_{WG_x} = -v_{BW}\sin\theta + v_{WG} = 0$, $v_{BG_y} = v_{BG} = v_{BW}\cos\theta$, $\sin\theta = \dfrac{v_{WG}}{v_{BW}} = \dfrac{2}{4}$, $\theta = 30°$, $v_{BG} = v_{BW}\cos 30° = 3.4$ km/h;

(2) $t = d/v_{BG} = 4/3.4 \approx 1.17$ h ≈ 70 min

Chapter 2

1. (1) (D)   (2) (C)   2. (C)   3. (A)   4. (B)   5. (C)
6. (D)   7. (1) (B)   (2) (C)   8. (1) (B)   (2) (D)   9. (C)
10. (1) (D)   (2) (E)   11. (1) (D)   (2) (C)
12. **Answer**: (1) $m_1\mu_k\cos\alpha + m_1\sin\alpha$; (2) $m_1\sin\alpha - m_1\mu_k\cos\alpha$;
(3) $(m_1\sin\alpha - m_1\mu_s\cos\alpha) < m_2 < (m_1\mu_s\cos\alpha + m_1\sin\alpha)$
13. **Answer**: (1) 40 J, 2.83 m/s; (2) 20 J, 3.46 m/s; (3) 60 J
14. **Answer**: (1) 0.062,5 J; (2) 0.177 m/s
15. **Answer**: 0.786
16. **Answer**: (1) 59.4 N; (2) ± 10 m/s
17. **Answer**: (1) 35 N; (2) 2.5 m/s
18. **Answer**: 2.46 $\hat{i}$ N
19. **Answer**: (1) 18.75 $\hat{i}$ N·s; (2) $-3.55 \times 10^{-3}\hat{j}$ N·s; (3) $7.5 \times 10^3\ \hat{i}$ N; (4) $12.95\ \hat{i} - 0.725\ \hat{j}$ N·s; (5) $89.3\ \hat{i} - 5.0\ \hat{j}$ m/s
20. **Answer**: (1) 1.25 kg; (2) $(1.5$ m/s$^3)t\ \hat{i}$; (3) 5.625 N
21. **Answer**: (1) $v(L) = \sqrt{\dfrac{g}{L}(L^2 - l^2)}$; (2) $t = \sqrt{\dfrac{L}{g}}\ln\dfrac{L + \sqrt{L^2 - l^2}}{l}$
22. **Answer**: (1) 125 J; (2) 50 J; (3) 66.7 J
23. **Answer**: (1) $x_m = x_0$; (2) $F(x) = 12\varepsilon\left(\dfrac{x_0^{12}}{x^{13}} - \dfrac{x_0^6}{x^7}\right)$; (3) $E_d = -U(x_0) = \varepsilon$
24. **Answer**: (1) $J_{mg} = \displaystyle\int_{t_1}^{t_2} m\boldsymbol{g}\,dt = \dfrac{\pi r}{v}m\boldsymbol{g}$; (2) $J_T = m\sqrt{4v^2 + \dfrac{\pi^2 r^2 g^2}{v^2}}$
25. **Answer**: $S = \dfrac{m}{m+M}R$
26. **Answer**: II lands a distance $3d$ from the starting point
27. **Answer**: $v_f = \dfrac{v_i R}{r} > v_i$

Chapter 3

1. (C)   2. (C)   3. (C)   4. (C)   5. (C)
6. (A)   7. (1) (A)   (2) (B)   8. (B)

Answers to Questions and Problems

9. **Answer**: (1) 16 rad, 250 rad; (2) 42 m; (3) 78 rad/s; (4) 54 rad/s
10. **Answer**: (1) 0.40 rad/s; (2) 6.60 rad
11. **Answer**: 6.0 rad/s²; 20 rad/s; 1.2 m/s
12. **Answer**: $v=\sqrt{\dfrac{2gh}{1+M/2m}}$; $\omega=\sqrt{\dfrac{2gh}{1+M/2m}}/R$
13. **Answer**: $I=\dfrac{1}{2}M(R_1^2+R_2^2)$
14. **Answer**: $v=\sqrt{\dfrac{2(m_B-\mu_k m_A)gd}{m_B+m_A+\dfrac{I}{R^2}}}$
15. **Answer**: $\dfrac{2}{3}g$; $\dfrac{1}{3}Mg$; $\sqrt{\dfrac{4}{3}gh}$
16. **Answer**: $\omega=\dfrac{I_A\omega_A+I_B\omega_B}{I_A+I_B}$
17. **Answer**: 1.6×10^4 J; 1.6×10^4 J; 2,000 W
18. **Answer**: 0.40 rad/s; 0.4 J
19. **Answer**: 7 rad/s; 0.01 J; 0.01 J
20. **Answer**: (1) $(-26.2 \text{ m/s})\,\hat{\boldsymbol{i}}$; (2) $(4.87 \text{ m/s}^2)\,\hat{\boldsymbol{i}}-(375 \text{ m/s}^2)\,\hat{\boldsymbol{j}}$; (3) 1.83 m
21. **Answer**: $F_R=15$ N; $F_L=27$ N
22. **Answer**: $v_{CM}=\sqrt{\dfrac{4}{3}Lg\sin\theta}=\sqrt{\dfrac{4}{3}gh}$
23. **Answer**: $\omega_z=253$ rad/s
24. **Answer**: $\omega=\dfrac{2m_0+m}{2m_0+4m}\omega_0$
25. **Answer**: $v=\sqrt{\dfrac{mgs^2}{(2m+m_0)l}}$; $a=\dfrac{2mgs}{(2m+m_0)l}$

Chapter 4

1. (B)   2. (A)   3. (B)   4. (A)   5. (B)
6. (B)   7. (A)   8. (B)   9. (C)   10. (A)   11. (B)

12. **Answer**: $\dfrac{Q}{4\pi\varepsilon_0 R}$

13. **Answer**: $\dfrac{\sigma}{8\varepsilon_0}[(z^2+R^2)^{1/2}-z]$

14. **Answer**: $2\pi\varepsilon_0\dfrac{R_1R_2}{R_2-R_1}V^2$

15. **Answer**: $C=\dfrac{\varepsilon_0 A}{d}\dfrac{\kappa_1+\kappa_2}{2}$

16. **Answer**: $\dfrac{q}{24\varepsilon_0}$

17. **Answer**: $F=\dfrac{Q^2}{2\varepsilon_0 A}$

18. **Answer**: $U = \dfrac{Q^2}{8\pi\varepsilon_0 R}$

19. **Answer**: 24 V

20. **Answer**: $2y(2z-1)\hat{i} - 2(y+x-2xz)\hat{j} + 4xy\hat{k}$

21. **Answer**:

Select small charge segment with amount of charge $dq = \lambda dx$, the magnitude of the electric field at point P is given by:

$$dE = \frac{1}{4\pi\varepsilon_0}\frac{dq}{(l+b-x)^2} = \frac{1}{4\pi\varepsilon_0}\frac{\lambda dx}{(l+b-x)^2} = \frac{A}{4\pi\varepsilon_0}\frac{xdx}{(l+b-x)^2}$$

The total electric field:

$$E = \int dE = \frac{A}{4\pi\varepsilon_0}\int_0^l \frac{xdx}{(l+b-x)^2} = \frac{A}{4\pi\varepsilon_0}\left[\frac{l}{b} - \ln\left(1+\frac{l}{b}\right)\right]$$

22. **Answer**:

(1) Select a spherical Gaussian surface of radius $r < a$

$$\oint_S \mathbf{E}\cdot d\mathbf{A} = E\oint_S dA = E(4\pi r^2) = \frac{q_{in}}{\varepsilon_0} = \frac{1}{\varepsilon_0}\rho\left(\frac{4}{3}\pi r^3\right) = \frac{q}{\varepsilon_0}\frac{r^3}{a^3}$$

$$E = \frac{1}{4\pi\varepsilon_0}\frac{q}{a^3}r \text{ (for } r<a\text{)}$$

(2) Select a spherical Gaussian surface of radius $a \leqslant r < b$

$$\oint_S \mathbf{E}\cdot d\mathbf{A} = E\oint_S dA = E(4\pi r^2) = \frac{q_{in}}{\varepsilon_0} = \frac{q}{\varepsilon_0}$$

$$E = \frac{1}{4\pi\varepsilon_0}\frac{q}{r^2} \text{ (for } a\leqslant r<b\text{)}$$

(3) $E = 0$ (for $b < r < c$)

(4) Select a spherical Gaussian surface outside the shell ($c \leqslant r$)

$$\oint_S \mathbf{E}\cdot d\mathbf{A} = E\oint_S dA = E(4\pi r^2) = \frac{q_{in}}{\varepsilon_0} = 0$$

so $E = 0$ (for $c \leqslant r$)

(5) Because the electric field is zero everywhere inside the conductor, the charge of the inner surface is $-q$ and that of the outer surface is 0.

23. **Answer**: By symmetry, we select a spherical Gaussian surface inside the shell ($a < r < b$)

$$\oint_S \mathbf{E}\cdot d\mathbf{A} = E\oint_S dA = E(4\pi r^2) = \frac{q_{in}}{\varepsilon_0}$$

$$q_{in} = q + \int\rho dV = q + \int_a^r \frac{A}{r}4\pi r^2 dr = q + 4\pi A\int_a^r r dr = q + 2\pi A(r^2 - a^2)$$

Compare the above two equations, we can conclude:

$$E = \frac{A}{2\varepsilon_0} + \frac{q - 2\pi A a^2}{4\pi r^2 \varepsilon_0}$$

When the constant A is satisfied with $q/2\pi a^2$, the electric field inside the shell is uniform and has the value:

$$E = \frac{A}{2\varepsilon_0} = \frac{q}{4\pi\varepsilon_0 a^2}$$

Answers to Questions and Problems

24. Answer:

(1) By symmetry, we select a cylindrical Gaussian surface of radius r outside the conduction shell and length L that is coaxial with the rod,

$$\oint_S \mathbf{E} \cdot d\mathbf{A} = \iint_{\text{side}} \mathbf{E} \cdot d\mathbf{A} + \iint_{\text{top\&bottom}} \mathbf{E} \cdot d\mathbf{A} = \iint_{\text{side}} \mathbf{E} \cdot d\mathbf{A} = E(2\pi rL) = \frac{q_{\text{in}}}{\varepsilon_0} = \frac{-q}{\varepsilon_0}$$

$E = -q/2\pi\varepsilon_0 Lr$, where "$-$" represents the direction of the electric field is radially inward.

(2) Because the electric field is zero everywhere inside the conducting shell, the charge of the inner surface is $-q$. According to the conservation of net charge, the charge of the outer surface is also $-q$.

(3) By symmetry, we select a cylindrical Gaussian surface of radius r and length L the electric field in the region between the shell and rod,

$$\oint_S \mathbf{E} \cdot d\mathbf{A} = \iint_{\text{side surface}} \mathbf{E} \cdot d\mathbf{A} = E(2\pi rL) = \frac{q_{\text{in}}}{\varepsilon_0} = \frac{q}{\varepsilon_0}$$

$E = +q/2\pi\varepsilon_0 Lr$, where "$+$" represents the direction of the electric field is radially outward.

25. Answer:

(1) The typical results for a parallel-plate capacitor are

$$C = \frac{\varepsilon_0 A}{d}, q = CV = \frac{\varepsilon_0 AV}{d}$$

When the separation is $2d$, C is decreasing, we get

$$q = \frac{\varepsilon_0 AV'}{2d}$$

The new potential difference is $V' = \frac{2d}{\varepsilon_0 A}q = \frac{2d}{\varepsilon_0 A}\frac{\varepsilon_0 A}{d}V = 2V$.

(2) The initial and final stored energies are

$$U_i = \frac{1}{2}CV^2 = \frac{\varepsilon_0 AV^2}{2d} \text{ and } U_f = \frac{1}{2}\frac{\varepsilon_0 A}{2d}(V')^2 = \frac{\varepsilon_0 AV^2}{d} \text{ respectively.}$$

(3) According to the work-energy principle, the work required to separate the plates is

$$W = U_f - U_i = \frac{\varepsilon_0 AV^2}{2d}$$

26. Answer:

(1) Choose a spherical shell segment with radius of r and thick dr, the charge element is

$$dq = (kr/R)4\pi r^2 dr = \frac{4\pi kr^3 dr}{R}$$

So the total charge of the sphere is

$$q(r) = \int_0^r \frac{4\pi kr^3}{R} dr = \frac{\pi k}{R}r^4, \text{ and } Q = q(R) = \pi kR^3$$

(2) Applying Gauss's law

$$\Phi_E = \oint E(r) dA = \begin{cases} q(r)/\varepsilon_0 & (r \leq R) \\ Q/\varepsilon_0 & (r \geq R) \end{cases}$$

· 189 ·

$$4\pi r^2 E(r) = \begin{cases} \dfrac{\pi k}{\varepsilon_0 R} r^4 & (r \leqslant R) \\ \dfrac{\pi k R^3}{\varepsilon_0} & (r \geqslant R) \end{cases}$$

So, the electric fields inside and outside of the sphere are

$$E(r) = \begin{cases} \dfrac{kr^2}{4\varepsilon_0 R} & (r \leqslant R) \\ \dfrac{kR^3}{4\varepsilon_0 r^2} & (r \geqslant R) \end{cases}$$

(3) The electric potential at the center of the sphere is

$$V(0) = \int_0^\infty E(r)\,dr = \int_0^R \dfrac{k}{4\varepsilon_0 R} r^2\,dr + \int_R^\infty \dfrac{kR^3}{4\varepsilon_0} \dfrac{dr}{r^2} = \dfrac{kR^2}{3\varepsilon_0}$$

27. Answer:

(1) The field from a large plate is perpendicular to the plate and uniform, $E = \sigma/2\varepsilon_0$. For regions outside the slab, it can be considered an infinite number of plates. We can find the equivalent surface density by considering the charge in the slab with an area A

$$Q_{\text{slab}} = \sigma_{\text{slab}} A = \rho_E A d, \text{ and } \sigma_{\text{slab}} = \rho_E d$$

The electric field to the left of the plate is

$$E_a = E_{\text{plate}} + E_{\text{slab}} = \sigma/2\varepsilon_0 + \sigma_{\text{slab}}/2\varepsilon_0 = (\sigma + \rho_E d)/2\varepsilon_0 \text{ (left)}$$

(2) The electric field to the right of the plate is

$$E_b = E_{\text{plate}} + E_{\text{slab}} = \sigma/2\varepsilon_0 + \sigma_{\text{slab}}/2\varepsilon_0 = (\sigma + \rho_E d)/2\varepsilon_0 \text{ (right)}$$

(3) To find the field inside the slab, we choose a cylinder for the Gaussian surface in Fig. 4-19. The cylinder is distance x inside the slab. Applying Gauss's law, we have

$$\oint E \cdot dA = \int_{\text{ends}} E \cdot dA + \int_{\text{side}} E \cdot dA = Q_{\text{en}}/\varepsilon_0$$

$$E_a A + E_c A + 0 = (\sigma A + \rho_E x A)/\varepsilon_0$$

$$E_c = \dfrac{\sigma + \rho_E(2x - d)}{2\varepsilon_0}$$

28. Answer:

We choose a ring of radius r and width dr for a small segment, with charge $dq = \sigma 2\pi r\,dr$. The potential of this element on the axis a distance x from the ring is

$$dV = \dfrac{dq}{4\pi\varepsilon_0 (x^2 + r^2)^{1/2}}$$

$$= \dfrac{\sigma 2\pi r\,dr}{4\pi\varepsilon_0 (x^2 + r^2)^{1/2}}$$

$$= \dfrac{ar^3\,dr}{2\varepsilon_0 (x^2 + r^2)^{1/2}}$$

The potential at points along the x axis is

$$V = \dfrac{a}{2\varepsilon_0} \int_0^R \dfrac{r^3\,dr}{(x^2 + r^2)^{1/2}}$$

$$= \dfrac{a}{2\varepsilon_0} \left[\dfrac{1}{3}(x^2 + R^2)^{1/2}(R^2 - 2x^2) - \dfrac{1}{3}x(-2x^2) \right]$$

$$= \frac{a}{6\varepsilon_0}\left[\frac{1}{3}(x^2+R^2)^{1/2}(R^2-2x^2)+2x^3\right]$$

29. Answer:

For a small angle θ, we have $\tan\theta \approx \theta$. If we consider a differential element a distance y from the small end, the capacitance of the element is

$$dC = \frac{\varepsilon dA}{(d+y\tan\theta)} \approx \varepsilon l dy/(d+y\theta)$$

where l is the length of a plate. The infinite number of elements is in parallel, so we find the total capacitance by integrating

$$C = \varepsilon l \int_0^l \frac{dy}{d+y\theta} = \frac{\varepsilon l}{\theta}\ln\left(\frac{d+l\theta}{d}\right) = \frac{\varepsilon l}{\theta}\ln\left(1+\frac{l\theta}{d}\right)$$

We use the expansion $\ln(1+x) \approx x - \frac{1}{2}x^2$ (for small x)

$$c \approx \frac{\varepsilon l}{\theta}\left[(l\theta/d) - \frac{1}{2}(l\theta/d)\right] = \frac{\varepsilon A}{d}\left[1 - \frac{1}{2}(\theta\sqrt{A/d})\right]$$

Chapter 5

1. (A)   2. (D)   3. (A)   4. (D)   5. (D)
6. (E)   7. (D)   8. (A)   9. (C)   10. (B)
11. (D)   12. (C)   13. (C)   14. (C)

15. **Answer:** -2.0 T

16. **Answer:** I

17. **Answer:** East

18. **Answer:** 0

19. **Answer:**

This is a typical problem in solving the magnetic field of steady current-carrying wire. One has to learn a lot of special results, such as wire, circular shape and arc shape objects.

The magnetic field at point O is consisting of four parts.

$$\boldsymbol{B}_0 = \boldsymbol{B}_1 + \boldsymbol{B}_2 + \boldsymbol{B}_3 + \boldsymbol{B}_4$$

$$\boldsymbol{B}_1 = \boldsymbol{B}_4 = 0$$

$$|B_2| = \frac{\mu_0 I}{8R}$$

$$|B_3| = \frac{\mu_0 I}{2\pi R}$$

So, the total magnetic field at point O is

$$B_0 = \frac{\mu_0 I}{8R} + \frac{\mu_0 I}{2\pi R} = \frac{\mu_0 I}{2R}\left(\frac{1}{4}+\frac{1}{\pi}\right)$$

The direction is into the paper.

20. **Answer:**

(1) $B = \dfrac{\mu_0 I}{4R_1} + \dfrac{\mu_0 I}{4R_2} = \dfrac{\mu_0 I}{4} \dfrac{R_1 + R_2}{R_1 R_2}$

Direction: into the page

(2) $\mu = IA = I\left(\dfrac{\pi R_1^2}{2} + \dfrac{\pi R_2^2}{2}\right) = \dfrac{\pi I}{2}(R_1^2 + R_2^2)$

Direction: into the page

21. **Answer**:

The magnetic field at the center of the ball can be obtained by the superposition of differential circular segment.

$$dI = In\,dl = I\dfrac{N}{\frac{1}{2}\pi R} R\,d\theta$$

$$dB = \dfrac{\mu_0}{2} \dfrac{y^2\,dI}{(x^2+y^2)^{3/2}} = \dfrac{\mu_0}{2} \dfrac{2N}{\pi R} \dfrac{y^2 I}{(x^2+y^2)^{3/2}} R\,d\theta$$

$$B = \int_0^{\pi/2} \dfrac{\mu_0 NI}{\pi} \dfrac{y^2}{(x^2+y^2)^{3/2}} d\theta = \int_0^{\pi/2} \dfrac{\mu_0 NI}{\pi R} \sin^2\theta\,d\theta = \dfrac{\mu_0 NI}{4R}$$

22. **Answer**:

(1) For the wire is in a circular shape, the net force by the magnet field exerted on the current loop is

$$F_{net} = 0$$

(2) According to the definition of torque, we can get

$$\mathbf{M} = \boldsymbol{\mu} \times \mathbf{B}$$

In this expression, $\boldsymbol{\mu}$ is the magnetic moment, and it is defined by

$$\boldsymbol{\mu} = \pi R^2\,I\mathbf{n}$$

The direction of torque is downward.

23. **Answer**:

The rotating charge is equivalent to a circular current. We choose a differential element of length dy a distance y form the axis of rotation. The charge on this element is $dq = (Q/l)dy$. The effective current of the element is $dI = dq/T = (\omega/2\pi)dq$.

Thus the magnetic moment of the element is

$$d\mu = A\,dI = (\pi y^2)(\omega/2\pi)dq = (\omega Q/2l)y^2\,dy$$

The total magnetic moment by integrating the magnetic moments of the differential elements

$$\mu = \int d\mu = (\omega Q/2l)\int_0^l y^2\,dy = (\omega Q/2l)(l^3/3) = \omega Q l^2/6$$

24. **Answer**:

(1) Figure(a) shows a view looking directly at the current. The sheet may be thought of as an infinite number of parallel wires. We choose a differential element of width dx a distance x form the center of the strip.

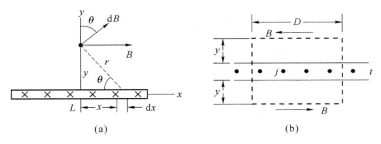

(a)   (b)

This current element is $dI = (I/L)dx$, and it produces a magnetic field in space $dB = (\mu_0/4\pi)2dI/r = (\mu_0 I/2\pi L)dx/r$.

We find the total field by integrating the x-components of the differential fields

$$B = \int \sin\theta\, dB = \int_{-L/2}^{L/2} \left(\frac{\mu_0 I}{2\pi L}\right)\left(\frac{y}{x^2+y^2}\right)dx$$

$$= \left(\frac{\mu_0 Iy}{2\pi L}\right)\left[\left(\frac{1}{y}\right)\arctan\left(\frac{x}{y}\right)\right]\bigg|_{-L/2}^{L/2} = \left(\frac{\mu_0 I}{\pi L}\right)\arctan\left(\frac{L}{2y}\right)$$

(2) If $y \gg L$ or $L/2y \ll 1$, the angle is small and equal to the tangent, so we have $\arctan(L/2y) \approx L/2y$ and $B \approx (\mu_0 I/\pi L)(L/2y) = (\mu_0/4\pi)(2L/y)$.

25. **Answer:**

Because the point C is along the line of the two straight segments of the wire, there is no magnetic field from these segments. The magnetic field at the point C is the sum of two fields

$$B = B_{\text{lower}} + B_{\text{upper}}$$

Each field is half that of a circular loop, with the field of the lower semicircle out the page and that of the upper semicircle into the page, so we subtract the two magnitudes

$$B = \frac{1}{2}\left(\mu_0 \frac{3}{4}I/2R\right) - \frac{1}{2}\left(\mu_0 \frac{1}{4}I/2R\right) = \mu_0 I/8R$$

The direction is out of the page.

26. **Answer:**

(1) The angle subtended by a side of the polygon at the center point P is

$$\theta = 2\pi/n$$

the length of a side is

$$L = 2R\sin(\theta/2) = 2R\sin(\pi/n)$$

the perpendicular distance from the center to the side is

$$D = R\cos(\theta/2) = R\cos(\pi/n)$$

the magnetic field of one side is

$$B_1 = \mu_0 I_0 L/2\pi D(L^2 + 4D^2)^{1/2}$$
$$= \mu_0 I_0 [2R\sin(\pi/n)]/2\pi[R\cos(\pi/n)][4R^2\sin^2(\pi/n) + 4R^2\cos^2(\pi/n)]^{1/2}$$
$$= \mu_0 I_0/2\pi[R\tan(\pi/n)]$$

Thus the field from the n sides is

$$B = nB_1 = [(\mu_0 I_0/2R)n\tan(\pi/n)]$$

into the page.

(2) If we let $n\to\infty$, the angle $\pi/n\to 0$, so $\tan(\pi/n)\to\pi/n$, thus we have $B\to(\mu_0 I_0/2\pi R)n$ $(\pi/n)=\mu_0 I_0/2R$, which is the expression for a circular loop.

Chapter 6

1. (C)   2. (D)   3. (D)   4. (B)   5. (A)
6. (A)   7. (D)   8. (C)   9. (C)   10. (C)
11. (B) 12. (B) 13. (D) 14. (B) 15. (C)
16. (D) 17. (D)
18. **Answer**: $0.004,5$ V/m
19. **Answer**: 80×10^{-6}; clockwise
20. **Answer**: counterclockwise
21. **Answer**: $\oint \boldsymbol{E}\cdot d\boldsymbol{A}=\dfrac{q}{\varepsilon_0}$; $\oint \boldsymbol{B}\cdot d\boldsymbol{A}=0$; $\oint \boldsymbol{E}\cdot d\boldsymbol{l}=-\dfrac{d\Phi_B}{dt}=-\int\dfrac{\partial \boldsymbol{B}}{\partial t}\cdot d\boldsymbol{A}$; $\oint \boldsymbol{B}\cdot d\boldsymbol{l}=\mu_0 I+\varepsilon_0\mu_0\dfrac{d\Phi_E}{dt}=\mu_0 I+\varepsilon_0\mu_0\int\dfrac{\partial \boldsymbol{E}}{\partial t}\cdot d\boldsymbol{A}$

22. **Answer**:

Applying Faraday's law, the induced emf can be obtained from the magnetic flux passing through the closed area.

(1) The total magnetic flux is

$$\Phi_B=\int_a^{a+L}d\Phi_B=\dfrac{\mu_0 ix}{2\pi}\ln\dfrac{a+L}{a}$$

So, the magnitude of induced emf is

$$\varepsilon=\dfrac{\mu_0 i}{2\pi}\ln\left(\dfrac{a+L}{a}\right)v=2.3\times 10^{-4}\text{ V}$$

(2) The current in the squared circuit is $i_l=\dfrac{\varepsilon}{R}=5.8\times 10^{-4}$ A

Direction: clockwise

23. **Answer**:

The distance between I_1 and I_2 is varying. To find the resultant force, a small segment is chosen, and the magnetic force exerted on current segment $I_2\,dl$ by I_1 is

$$|dF_{21}|=|I_2 d\boldsymbol{l}\times\boldsymbol{B}_1|=I_2 dl\dfrac{\mu_0 I_1}{2\pi r}$$

The resultant force acting on current I_2 is obtained by integrating

$$F_{12}=\dfrac{\mu_0 I_1 I_2}{2\pi}\int_0^L\dfrac{dx}{d+x\cos\theta}=\dfrac{\mu_0 I_1 I_2}{2\pi\cos\theta}\ln\left(\dfrac{d+L\cos\theta}{d}\right)$$

Direction is perpendicular and upward to the metal rod.

24. **Answer**:

Choose direction of loop is clockwise.

(1) For uniform magnetic field

$$\Phi_B = B\frac{1}{2}x^2\tan\theta = \frac{1}{2}Bv^2t^2\tan\theta$$

$$E = -\frac{d\Phi_B}{dt} = -Bv^2t\tan\theta$$

The direction of induced emf is clockwise.

(2) The magnetic flux is

$$\Phi_B = \int_0^x K\xi\cos\omega t \xi\tan\theta d\xi = \int_0^x K\xi^2 d\xi(\tan\theta\cos\omega t) = \frac{1}{3}Kx^3\tan\theta\cos\omega t$$

$$E = -\frac{d\Phi_B}{dt} = -Kx^2\frac{dx}{dt}\tan\theta\cos\omega t + \frac{1}{3}Kx^3\omega\tan\theta\sin\omega t$$

$$= Kv^3\tan\theta\left(\frac{1}{3}\omega t^3\sin\omega t - t^2\cos\omega t\right)$$

The positive sign of emf is set to be the direction of clockwise.

25. **Answer:**

(1) The magnetic field is expressed as

$$B(x) = \frac{\mu_0 i}{2\pi x}$$

The total magnetic flux is obtained by

$$\Phi_B = \int_c^{b+c} B(x)a dx = \frac{\mu_0 ia}{2\pi}\int_c^{b+c}\frac{dx}{x} = \frac{\mu_0 ia}{2\pi}\ln\frac{b+c}{c}$$

(2) The loop is moving. Assume the coordinate of the lower edge of the loop is $x'(t) = c + vt$, the coordinate of the upper edge is $x(t) + b$, then the flux of the loop is

$$\Phi(t) = \frac{\mu_0 ia}{2\pi}\ln\frac{x'(t)+b}{x'(t)}$$

The induced emf and current can be expressed as following

$$\varepsilon = -\frac{d\Phi}{dt} = -\frac{\mu_0 ia}{2\pi}\times\frac{x'}{(x'+b)}\times\frac{[x'-(x'+b)]}{x^2}\times\frac{dx'}{dt} = \frac{\mu_0 iabv}{2\pi x'(x'+b)}$$

$$i_{ind} = \frac{\varepsilon}{R} = \frac{\mu_0 iabv}{2\pi R}\frac{1}{x'(x'+b)} = \frac{\mu_0 iabv}{2\pi R}\frac{1}{(c+vt)(c+b+vt)}$$

26. **Answer:**

We find the mutual inductance by finding the linkage of the magnetic field of loop 1 with loop 2. Because the loops are small, we can assume the field from loop 1 at loop 2 is constant and equal to the magnitude at the center of the loop $B_1 = \mu_0 I_1 r_1^2/2(r_1^2+l^2)^{3/2}$. Because $r_1 \ll l$, this becomes $B_1 = \mu_0 I_1 r_1^2/2l^3$. The flux linking with loop 2 is $\Phi_{21} = B_1\pi r_2^2\cos\theta$, so the mutual inductance is $M = \Phi_{21}/I_1 = (\mu_0\pi r_1^2 r_2^2/2l^3)\cos\theta$.

27. **Answer:**

(1) We choose the direction of current as the z-axis. The electric field inside and on the surface from the current density is

$$E = J/\sigma = (I/A\sigma)k = (I/\pi r^2\sigma)k = (IR/L)k$$

where R is the resistance of a length L.

(2) From the cylindrical symmetry, the magnetic field will be circular, centered on the

· 195 ·

axis of the wire. Choose a circular path with radius $r_1 > r$ to apply Ampère's law

$$\oint B \cdot ds = \mu_0 I_{enclosed}$$

$$B 2\pi r = \mu_0 I$$

which gives $B = \dfrac{\mu_0 I}{2\pi r_1}$, for $r_1 > r$.

Chapter 7

1. (A)   2. (B)   3. (C)   4. (B)
5. **Answer**: 567.42 kPa
6. **Answer**: 1,900
7. **Answer**: 1.13
8. **Answer**: 2 : 1
9. **Answer**:

(1) From Fig. 7-4, the speed distribution function can be expressed as following

$$f(v) = \begin{cases} \dfrac{a}{v_0} v, & 0 < v < v_0 \\ a, & v_0 < v < 2v_0 \\ 0, & v > 2v_0 \end{cases}$$

Applying for the normalization condition, we can see that

$$\int_0^\infty f(v) dv = \int_0^{v_0} \dfrac{a}{v_0} v\, dv + \int_{v_0}^{2v_0} a\, dv + 0 = 1$$

$$\dfrac{a}{2v_0} v_0^2 + a v_0 = 1$$

$$a = \dfrac{2}{3v_0}$$

(2) The number of molecules can be obtained as follows

$$\Delta N_{v > v_0} = \int_{v_0}^\infty dN = \int_{v_0}^\infty N f(v) dv = \int_{v_0}^{2v_0} N a\, dv = \dfrac{2N}{3v_0} v_0 = \dfrac{2}{3} N$$

$$\Delta N_{v < v_0} = \int_0^{v_0} dN = \int_0^{v_0} N f(v) dv = \int_0^{v_0} N \dfrac{a}{v_0} v\, dv = \dfrac{Na}{v_0} \cdot \dfrac{v_0^2}{2} = \dfrac{1}{3} N$$

(3) The average speed is

$$\bar{v} = \int_0^\infty v f(v) dv = \int_0^{v_0} \dfrac{a}{v_0} v^2 dv + \int_{v_0}^{2v_0} a v\, dv = \dfrac{2}{9} v_0 + v_0 = \dfrac{11}{9} v_0$$

Chapter 8

1. (C)   2. (C)   3. (A)   4. (D)   5. (C)   6. (D)
7. **Answer**: 5.25×10^{-3}
8. **Answer**: 5,320
9. **Answer**: 0

Answers to Questions and Problems

10. **Answer:** 765

11. **Answer:** 192

12. **Answer:** 2.5×10^4

13. **Answer:**

(1) Use the first law of thermodynamics to find $\Delta U_{ca} = U_a - U_c$.
$$\Delta U_{ca} = -\Delta U_{ac} = -(Q_{\text{curve } a \to c} - W_{\text{curve } a \to c}) = -(-85 \text{ J} - (-55 \text{ J})) = 30 \text{ J}$$

(2) Use the first law of thermodynamics to find Q_{cda}.
$$\Delta U_{ca} = Q_{cda} - W_{cda} \Rightarrow \Delta U_{ca} + W_{cda} = 30 \text{ J} + 38 \text{ J} = 68 \text{ J}$$

(3) Since the work along path bc is zero,
$$W_{abc} = W_{ab} = p_a(V_b - V_a) = 2.2 p_d(V_c - V_d) = 2.2(-W_{cda}) = 2.2(-38 \text{ J}) = -84 \text{ J}$$

(4) Use the first law of thermodynamics to find Q_{abc}.
$$Q_{abc} = \Delta E_{ac} + W_{abc} = -30 \text{ J} - 84 \text{ J} = -114 \text{ J}$$

(5) Since $U_a - U_b = 15 \text{ J} \Rightarrow U_b = U_a - 15 \text{ J}$, we have the following
$$\Delta U_{bc} = U_c - U_b = U_c - (U_a - 15 \text{ J}) = U_c - U_a + 15 \text{ J} = -30 \text{ J} + 15 \text{ J} = -15 \text{ J}$$

Use the first law of thermodynamics to find Q_{bc}.
$$Q_{bc} = \Delta E_{bc} + W_{bc} = -15 \text{ J} + 0 = -15 \text{ J}$$

14. **Answer:**

(1) Leg ba is an isobaric expansion
$$W_{ba} > 0, \Delta U_{ba} > 0, Q_{ba} = \Delta U_{ba} + W_{ba} > 0.$$

Leg ad is an isovolumetric reduction in pressure
$$W_{ad} > 0, \Delta U_{ad} < 0, Q_{ad} = \Delta U_{ad} + W_{ad} < 0.$$

Leg dc is an isobaric compression
$$W_{dc} < 0, \Delta U_{dc} < 0, Q_{dc} = \Delta U_{dc} + W_{dc} < 0.$$

Leg cb is an isovolumetric expansion in pressure
$$W_{cb} = 0, \Delta U_{cb} > 0, Q_{cb} = \Delta U_{cb} + W_{cb} > 0.$$

(2) $W_{\text{cycle}} = W_{cba} + W_{adc} = W_{ba} - W_{cda} = W_{ba} - 38 \text{ J}$
$$W_{ba} = p_a(V_a - V_b) = 2.2 p_d(V_d - V_c) = 2.2 W_{cda} = 2.2 \times 38 \text{ J} = 83.6 \text{ J}$$
$$W_{\text{cycle}} = 83.6 \text{ J} - 38 \text{ J} = 45.6 \text{ J}$$

(3) Since the process is a cycle, the internal energy does not change.
$$\Delta U_{\text{cycle}} = 0$$

(4) Use the first law of thermodynamics, applied to the entire cycle.
$$Q_{\text{cycle}} = \Delta U_{\text{cycle}} + W_{\text{cycle}} = 0 + 45.6 \text{ J} = 45.6 \text{ J}$$

(5) From problem 13, we have $Q_{cba} = -Q_{abc} = 114 \text{ J}$. This is the intake heat. Therefore, the efficiency is
$$e = \frac{W_{\text{cycle}}}{Q_{\text{cycle}}} \times 100\% = \frac{45.6 \text{ J}}{114 \text{ J}} \times 100\% = 40\%$$

Chapter 9

1. (B)   2. (D)   3. (A)   4. (C)   5. (B)

6. (C)   7. (D)   8. (D)
9. **Answer**: 0.825
10. **Answer**: 1
11. **Answer**: 67
12. **Answer**: 3.5%
13. **Answer**: 113.04
14. **Answer**: 1.2×10^{20} J
15. **Answer**: 27
16. **Answer**:

Because process ab is isothermal, $\Delta U_{ab} = 0$, and
$$Q_{ab} = W_{ab} = nRT_H \ln(V_b/V_a)$$
Because process bc is constant volume, we have
$$Q_{bc} = \Delta U_{bc} = nC_{V,m}(T_L - T_H) = n\frac{3}{2}R(T_L - T_H)$$
$$W_{bc} = 0$$
Because process cd is isothermal, $\Delta U_{cd} = 0$, and
$$Q_{cd} = W_{cd} = nRT_L \ln(V_d/V_c) = nRT_L \ln(V_a/V_b) = -nRT_L \ln(V_b/V_a)$$
Because process da is constant volume, we have
$$Q_{da} = \Delta U_{da} = nC_{V,m}(T_H - T_L) = n\frac{3}{2}R(T_H - T_L)$$
$$W_{da} = 0$$
The net work done by the cycle is
$$W = W_{ab} + W_{cd} = nRT_H \ln(V_b/V_a) - nRT_L \ln(V_b/V_a) = nR(T_H - T_L)\ln(V_b/V_a)$$
The heat added to the system is
$$Q_{in} = Q_{ab} + Q_{da} = nRT_H \ln(V_b/V_a) + n\frac{3}{2}R(T_H - T_L)$$
The efficiency is
$$e_{Stirling} = W/Q_{in} = nR(T_H - T_L)\ln(V_b/V_a) / \left[nRT_H \ln(V_b/V_a) + n\frac{3}{2}R(T_H - T_L) \right]$$
$$= (T_H - T_L)\ln(V_b/V_a) / \left[T_H \ln(V_b/V_a) + \frac{3}{2}(T_H - T_L) \right]$$
The efficiency of a Carnot cycle is
$$e_{Carnot} = (T_H - T_L)/T_H$$
We rearrange the expression for the Stirling cycle
$$e_{Stirling} = [(T_H - T_L)/T_H]\left\{ 1 / \left[1 + \frac{3}{2}(T_H - T_L)/T_H \ln(V_b/V_a) \right] \right\}$$
Because the denominator of the second factor is greater than 1, we can conclude that
$$e_{Stirling} < e_{Carnot}$$

17. **Answer**:

For a monatomic gas $C_{p,m} = 5R/2, C_{V,m} = 3/2, \gamma = C_{p,m}/C_{V,m} = 5/3$
(1) We find the pressure at b from the ideal gas equation
$$(p_b/p_a)/(V_b/V_a)$$
$$(p_b/101.325 \text{ kPa}) \times 2.5 = 1, \text{ so } p_b = 40.53 \text{ kPa}$$

Answers to Questions and Problems

For the adiabatic compression, we have
$$p_a/V_a^\gamma = p_c V_c^\gamma, \text{ or } p_c/p_a = (V_a/V_c)^\gamma$$
$$p_c/101.325 \text{ kPa} = (1/2.5)^{5/3}, \text{ so } p_c = 21.99 \text{ kPa}$$

(2) We find the temperature at c from the ideal gas equation
$$p_c/p_a/(V_c/V_a) = T_c/T_a$$
$$(21.99 \text{ kPa}/101.325 \text{ kPa}) \times 2.5 = T_c/273 \text{ K}, \text{ so } T_c = 148 \text{ K}$$

① For process ab, the isothermal expression, we have
$$\Delta U_{ab} = 0$$
$$Q_{ab} = W_{ab} = nRT_a \ln(V_b/V_a) = 1 \text{ mol} \times 8.315 \text{ J/(mol·K)} \times 273 \text{ K} \ln 2.5$$
$$= 2.08 \times 10^3 \text{ J}$$
$$\Delta S_{ab} = Q_{ab}/T_a = (2.08 \times 10^3 \text{ J})/273 \text{ K} = 7.62 \text{ J/K}$$

For process bc, at constant volume, we have
$$\Delta W_{ab} = 0$$
$$Q_{bc} = \Delta U_{bc} = nC_V(T_c - T_b) = 1 \text{ mol} \times \frac{3}{2} \times 8.315 \text{ J/(mol·k)} \times (148 \text{ K} - 273 \text{ K}) \times \ln(2.5)$$
$$= -1.56 \times 10^3 \text{ J}$$
$$\Delta S_{bc} = nC_V \ln(T_c/T_b) = 1 \text{ mol} \times \frac{3}{2} \times 8.315 \text{ J/(mol·k)} \times \ln(148 \text{ K}/273 \text{ K}) = -7.64 \text{ J/K}$$

For the process ca, the adiabatic compression, we have
$$Q_{ca} = 0$$
$$W_{ca} = -\Delta U_{ca} = -nC_{V,m}(T_a - T_c) = -1 \text{ mol} \times \frac{3}{2} \times (273 \text{ K} - 148 \text{ K})$$
$$= -1.56 \times 10^3 \text{ J}$$
$$\Delta S_{ca} = 0$$

② The efficiency of the cycle is
$$e = W/Q_H = (W_{ab} + W_{ca})/Q_{ab} = (2.08 \times 10^3 \text{ J} - 1.56 \times 10^3 \text{ J})/(2.08 \times 10^3 \text{ J})$$
$$= 0.25 = 25\%$$

Chapter 10

1. (D)   2. (C)   3. (D)   4. (D)   5. (C)

6. **Answer:** $-A\omega^2 \cos \omega t$

7. **Answer:** $x = 0.05 \cos 10t$

8. **Answer:** 39 rad/s

9. **Answer:** $\frac{1}{2\pi}\sqrt{\frac{2k}{m}}$

10. **Answer:** $4A$

11. **Answer:** 49.7

12. **Answer:** 1, 2, 2

13. **Answer:**

(1) If we apply a force F to stretch the springs, the total displacement Δx is the sum of the displacements of the two springs

$$\Delta x = \Delta x_1 + \Delta x_2$$

The effective spring constant is defined from $F = -k_{\text{eff}} \Delta x$. Because they are in series, the force must be the same in each spring

$$F_1 = F_2 = F = -k_1 \Delta x_1 = -k_2 \Delta x_2$$

Then $\Delta x = \Delta x_1 + \Delta x_2$ becomes

$$-F/k_{\text{eff}} = -(F/k_1) - (F/k_2) = 0, \text{ or } 1/k_{\text{eff}} = 1/k_1 + 1/k_2$$

For the period we have

$$T = 2\pi (m/k_{\text{eff}})^{1/2} = 2\pi \{m[(1/k_1) + (1/k_2)]\}^{1/2}$$

(2) In the equilibrium position, we have

$$F_{\text{net}} = F_{20} - F_{10} = 0, \text{ or } F_{10} = F_{20}$$

When the object is moved to the right a distance x, we have

$$F_{\text{net}} = F_{20} - k_2 x - (F_{10} + k_1 x) = -(k_1 + k_2) x$$

The effective spring constant is $k_{\text{eff}} = k_1 + k_2$, so the period is

$$T = 2\pi (m/k_{\text{eff}})^{1/2} = 2\pi [m/(k_1 + k_2)]^{1/2}$$

14. **Answer**:

The angular frequency is

$$\omega = 2\pi f = (k/m)^{1/2} = [(250 \text{ N/m})/0.380 \text{ kg}]^{1/2} = 25.6 \text{ s}^{-1}$$

The period is $T = 2\pi/\omega = 2\pi/25.6 \text{ s}^{-1} = 0.245 \text{ s}$.

(1) For a general displacement, we have

$$x = A\cos(\omega t + \phi); v = -A\sin(\omega t + \phi)$$

Because $v = v_{\text{max}} = A\omega$, when $t_0 = 0.110 \text{ s}$, we have

$\sin(25.6 \text{ s}^{-1} \times 0.110 \text{ s} + \phi) = -1$, or $2.82 + \phi = 3\pi/2$, which gives $\phi = 1.89 \text{ rad} = 108°$

Note that this also gives $x = 0$

Thus the equation for the motion is

$$x = (12.0 \text{ cm}) \cos[(25.6 \text{ s}^{-1})t + 1.89 \text{ rad}]$$

(2) Because the mass passes through the equilibrium position toward positive x at t_0, it will reach the maximum length, $x = -12.0 \text{ cm}, \frac{3}{4}T$ later

$$t_{\text{max1}} = t_0 + \frac{3}{4}T = 0.110 \text{ s} + \frac{3}{4} \times 0.245 \text{ s} = 0.294 \text{ s}$$

It will be at this position at interval of T

$$t_{\text{max}} = 0.294 \text{ s}, 0.539 \text{ s}, 0.784 \text{ s}, \cdots$$

The mass will reach the minimum length, $x = 12.0 \text{ cm}, \frac{1}{4}T$ after t_0

$$t_{\text{min1}} = t_0 + \frac{1}{4}T = 0.110 \text{ s} + \frac{1}{4} \times 0.245 \text{ s} = 0.171 \text{ s}$$

It will be at this position at intervals of T

$$t_{\text{min}} = 0.171 \text{ s}, 0.416 \text{ s}, 0.661 \text{ s}, \cdots$$

Answers to Questions and Problems

(3) The displacement at $t=0$ is
$$x=(12.0 \text{ cm})\cos(0+1.89 \text{ rad})=-3.77 \text{ cm}$$

(4) Because the net force produces the SHM, the force in the spring at $t=0$ is
$$F=-kx+mg=(-250 \text{ N/m})\times(-0.037,7 \text{ m})+0.380 \text{ kg}\times 9.80 \text{ m/s}^2=13.1 \text{ N}$$

(5) The maximum speed is
$$v_{\max}=A\omega=0.120 \text{ m}\times 25.6 \text{ s}^{-1}=3.07 \text{ m/s}$$

Because the initial phase is less than 180°, the maximum speed is reached at $t=t_0=0.110$ s

15. **Answer:**

When a mass is a distance x from the center of the Earth, there will be a gravitational force only from the mass within a sphere of radius x. This mass is
$$M'=\left(M_E/\frac{4}{3}\pi r_E^3\right)\frac{4}{3}\pi x^3=M_E x^3/r_E^3$$

The force will be toward the center, opposite to x
$$F=-GM'm/x^2=-G(M_E x^3/r_E^3)m/x^2=-(GM_E m/r_E^3)x$$

Thus we see that the restoring force is proportional to the displacement, so the motion will be simple harmonic, with
$$k_{\text{eff}}=GM_E m/r_E^3$$

The apple will take one period to return to the initial location
$$T=2\pi(m/k_{\text{eff}})^{1/2}=2\pi[m/(GM_E m/r_E^3)]^{1/2}=2\pi(r_E^3/GM_E)^{1/2}$$
$$=2\pi[(6.38\times 10^6 \text{ m})^3/(6.67\times 10^{-11} \text{ N}\cdot\text{m}^2/\text{kg}^2)(5.98\times 10^{24} \text{ kg})]^{1/2}$$
$$=5.07\times 10^3 \text{ s}=84.5 \text{ min}$$

Chapter 11

1. (D) 2. (C) 3. (D) 4. (C) 5. (E)
6. (C) 7. (D) 8. (B)

9. **Answer:** 4

10. **Answer:** 1.01

11. **Answer:** 526

12. **Answer:** 269 or 259

13. **Answer:**

(1) For the sum of the two waves we have
$$y=y_1+y_2=A\sin(kx-\omega t)+A\sin(kx-\omega t+\phi)=2A\cos\left(\frac{1}{2}\phi\right)\sin\left(kx-\omega t+\frac{1}{2}\phi\right)$$

(2) The amplitude is the coefficient of the sine function: $2A\cos\left(\frac{1}{2}\phi\right)$. The variation in

x, and t is purely sinusoidal.

(3) If $\phi=0, 2\pi, 4\pi$; we can get $\cos\left(\frac{1}{2}\phi\right)=\pm 1$, thus the amplitude is maximum and we have complete constructive interference.

If $\phi=\pi, 3\pi, 5\pi$; we can get $\cos\left(\frac{1}{2}\phi\right)=0$, thus the amplitude is zero and we have destructive interference.

14. **Answer:**

(1) Because two loops are a wavelength, we have $\lambda=\frac{2}{3}L=\frac{2}{3}\times 1.80 \text{ m}=1.20 \text{ m}$

The wave characteristics are
$$k=2\pi/\lambda=2\pi/1.20 \text{ m}=5.24 \text{ m}^{-1}$$
$$\omega=2\pi f=2\pi\times 120 \text{ Hz}=240\pi \text{ s}^{-1}$$

The wave function is
$$y=A\sin(kx)\cos(\omega t)=(6.0 \text{ cm})\sin[(5.24 \text{ m}^{-1})x]\cos[(240\pi \text{ s}^{-1})t]$$

(2) The two waves traveling in opposite directions will have half the amplitude of the standing wave, but the same value for k and ω
$$y_1=(3.0 \text{ cm})\sin[(5.24 \text{ m}^{-1})x-(240\pi \text{ s}^{-1})t]$$
$$y_2=(3.0 \text{ cm})\sin[(5.24 \text{ m}^{-1})x+(240\pi \text{ s}^{-1})t]$$

15. **Answer:**

(1) The string will resonate at only certain frequencies, determined by the wave speed v on the string and the length of L of the string. These resonant frequencies are
$$f=n\frac{v}{2L} \text{ for } n=1,2,3,\cdots$$

To set up the fourth harmonic (for which $n=4$), we need to adjust the right side of this equation, with $n=4$, so that the left side equals the frequency of the vibrator (120 Hz). The tension in the cord is equal to the mass m of suspended object, thus we can get
$$v=\sqrt{\frac{\tau}{\mu}}=\sqrt{\frac{mg}{\mu}}$$

So, the mass of the object can be determined.
$$m=\frac{4L^2 f^2 \mu}{n^2 g}=\frac{4\times(1.2 \text{ m})^2\times(120 \text{ Hz})^2\times 0.001,6 \text{ kg/m}}{4^2\times 9.8 \text{ m/s}^2}=0.846 \text{ kg}\approx 0.85 \text{ kg}$$

(2) If we choose $m=1.00$ kg, we find that $n=3.7$. This means that there will be no standing wave on the string. Any oscillation of the string will be small, perhaps even imperceptible.

Chapter 12

1. (A)　　2. (D)　　3. (C)　　4. (C)

Answers to Questions and Problems

5. **Answer:** 2; 0.03 L/c

6. **Answer:** 155 nm; 310 nm

7. **Answer:** (b); (c); (a); (d)

8. **Answer:** $2.39E_0 \cos(\omega t + 8.8°)$

9. **Answer:**

For the left-hand end and right-hand end, the corresponding fringes are dark, and there are 4 dark fringes between. So for left-hand end, we have

$$2e_m + \frac{\lambda}{2} = (2m+1)\frac{\lambda}{2}$$

where $e_m = L_L$. And for right-hand end, we have

$$2e_{m+5} + \frac{\lambda}{2} = [2(m+5)+1]\frac{\lambda}{2}$$

where $e_{m+5} = L_R$.

Therefore

$$2(L_R - L_L) = 5\lambda, \Delta L = \frac{5\lambda}{2} = 2.5 \times 632.8 \text{ nm} = 1.582 \text{ μm}$$

10. **Answer:**

(1) For the bright fringes of Newton's ring, the radii of the rings are

$$r_m = \sqrt{\left(m - \frac{1}{2}\right)\lambda R}, \quad m = 1, 2, 3, \cdots$$

So

$$\Delta r = r_{m+1} - r_m = \left(\sqrt{m+1-\frac{1}{2}} - \sqrt{m-\frac{1}{2}}\right)\sqrt{\lambda R}$$

$$\xrightarrow{m \gg 1} \sqrt{m}\left[\left(1+\frac{1}{2m}\right)^{1/2} - \left(1-\frac{1}{2m}\right)^{1/2}\right]\sqrt{\lambda R}$$

$$\approx \sqrt{m\lambda R}\left(1 + \frac{1}{4m} - 1 + \frac{1}{4m}\right) = \frac{1}{2}\sqrt{\lambda R/m}$$

(2) The area of m-th order bright ring is $A_m = \pi r_m^2$

So the area between adjacent bright rings is

$$A = A_{m+1} - A_m = \pi(r_{m+1}^2 - r_m^2) = \pi\left[\left(m+1-\frac{1}{2}\right) - \left(m-\frac{1}{2}\right)\right]\lambda R = \pi \lambda R$$

11. **Answer:**

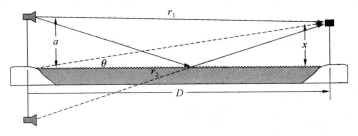

The apparatus is like Lloyd's mirror. Because $D \gg x, a$, the difference of optical path length is

$$r_2 - r_1 + \frac{\lambda}{2} = 2a\sin\theta + \frac{\lambda}{2}$$

where $\lambda/2$ comes from the reflection at the surface of the water.

The receiver gets the maximum signals so that

$$2a\sin\theta + \frac{\lambda}{2} = m\lambda$$

For the reason of $D \gg x$, $\sin\theta \approx \tan\theta = \frac{x}{D}$.

$$2a\frac{x}{D} = m\lambda - \frac{\lambda}{2}$$

$$x = \left(m - \frac{1}{2}\right)\frac{\lambda}{2a}D$$

Take the x as small as possible, we take $m = 1$.

$$x = \frac{\lambda}{4a}D$$

12. **Answer:**

When the mirror M_2 moves a distance of ΔL, the number of fringes N that shift across the vision field will satisfy following equation:

$$\Delta L = N\frac{\lambda}{2}$$

$$\lambda = \frac{2\Delta L}{N} = \frac{2 \times 233 \ \mu m}{792} = 0.588 \ \mu m = 588 \ nm$$

Chapter 13

1. (B)   2. (D)   3. (C)   4. (A)   5. (D)

6. **Answer:** 4

7. **Answer:** The 2nd order maximum

8. **Answer:** (1) red; (2) violet

9. **Answer:** diminish

10. **Answer:** (1) to the left; (2) less than

11. **Answer:** (1) less than; (2) greater than; (3) greater than

12. **Answer:** (1) A; (2) to the left

13. **Answer:** (1) left pair; (2) to the right

14. **Answer:** (1) $N = 6$

(2)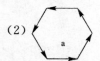

Answers to Questions and Problems

(3)

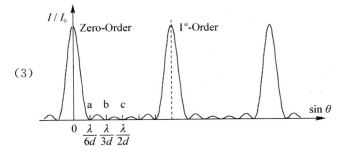

15. **Answer:**

(1) Suppose the order of maximum for $\sin\theta=0.2$ is m. Then

$$\sin\theta_m = m\frac{\lambda}{d} = 0.2$$

$$\sin\theta_{m+1} = (m+1)\frac{\lambda}{d} = 0.3$$

$$\frac{m}{m+1} = \frac{2}{3}, \quad m=2, m+1=3, d=2\frac{600\text{ nm}}{0.2}=6.00\ \mu\text{m}$$

The position for the 2nd order maximum

$$\theta_2 = \arcsin 0.2 = 11.5°, \theta_2 = \arcsin 0.3 = 17.5°$$

$$\Delta\theta = 6.0°$$

(2) The 4th order maximum is missing. So

$$\frac{d}{a}=4, a=\frac{d}{4}=\frac{6.0\ \mu\text{m}}{4}=1.5\ \mu\text{m}$$

(3) The highest order that can be seen on the viewing screen

$$|\sin 90°| > m_{max}\frac{\lambda}{d}, m_{max}=9$$

The orders of maxima that can be viewed on the screen are $0, \pm1, \pm2, \pm3, \pm5, \pm6, \pm7, \pm9$. The $\pm4, \pm8$ orders will be missing.

Chapter 14

1. (D)   2. (B)   3. (D)   4. (A)   5. (B)

6. **Answer:** $I = 0.5 \times (\cos 60°)^4 = 1/32 = 3.1\%$

7. **Answer:**

(1) Two of polarizing sheets are least required for rotating the plane of polarization through 90° like the following figure.

Assume the original light beam is vertically polarized, and the first polarizing sheet is orientated by θ with the vertical direction.

The light beam which gets through the second polarizing sheet is

$$A^2\cos^2\theta\cos^2(90°-\theta) = \frac{A^2}{4}4\sin^2\theta\cos^2\theta = \frac{A^2}{4}\sin^2 2\theta$$

· 205 ·

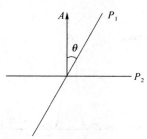

The maximum transmitted intensity is then 25% with $\theta=45°$.

(2) According to the fact that the transmitted intensity I_{out} is to be more than 60% of the original intensity

$$\cos^{2n}\frac{90°}{n}\geqslant 60\%=\frac{3}{5}$$

For $n=2$, $I_{out}=25\%\ I_{in}$. For $n=3$, $I_{out}=42\%\ I_{in}$; $n=4$, $I_{out}=53.1\%\ I_{in}$; $n=5$, $I_{out}=60.5\%\ I_{in}$.

So the minimum number of sheets required is 5.

8. Answer:

A partially polarized beam can be considered as a mixture of specific amount of natural light and polarized light, as shown in Fig. 14-6. According to the question, we have

$$I_x=\frac{I_{nature}}{2}, I_y=\frac{I_{nature}}{2}+I_{linear}, I_y=5I_x, I_x+I_y=I_0$$

So

$$\frac{I_{nature}}{2}+I_{linear}=5\frac{I_{nature}}{2}, \frac{I_{nature}}{2}+I_{linear}+\frac{I_{nature}}{2}=I_0$$

$$I_{linear}=\frac{2}{3}I_0$$

9. Answer:

(1) Because the sum of the angle of reflection $\theta_i=\theta_p$ and refraction θ_r when a beam is incident at the polarizing angle is 90°.

So (2) $\theta_p=90°-31.8°=58.2°$

Begin with the Brewster's law

$$n=\tan\theta_p=\tan 58.2°=1.61$$

10. Answer:

$$\sin\theta_c=\sin 52°=\frac{n_2}{n_1}=\tan\theta_p=0.788$$

$$\theta_p=38.2°$$

11. Answer:

The linearly polarized light has the property that the phase difference between the components along e-axis and o-axis is either 0 or π. Therefore we have

$$\frac{2\pi}{\lambda}(n_e-n_o)d=\pi$$

$$d=\frac{\lambda}{2(n_e-n_o)}=\frac{0.525\ \mu m}{2\times 0.022}=11.9\ \mu m$$

Answers to Questions and Problems

12. **Answer:**

For the calcite crystal in Fig. 14-21. The o-wave surface and e-wave surface is shown in the following figure.

So the Ray x is e-ray and Ray y is o-ray, their polarization states are shown in the following figure.

$$\sin \theta_i = n_o \sin \theta_{ro} = n_e \sin \theta_{re}$$

$$\sin \theta_{ro} = \frac{\sin \theta_i}{n_o} = \frac{\sin 38.8°}{1.658} = 0.378$$

$$\sin \theta_{re} = \frac{\sin \theta_i}{n_e} = \frac{\sin 38.8°}{1.486} = 0.422$$

The distance of A and B:

$$t(\tan \theta_{re} - \tan \theta_{ro}) = 1.12 \text{ cm} \times (0.465 - 0.408) = 0.063,8 \text{ cm}$$

The perpendicular distance between the two emerging rays x and y:

$$0.063,8 \text{ cm} \times \cos \theta_i = 0.063,8 \text{ cm} \times \cos 38.8° = 0.049,8 \text{ cm}$$

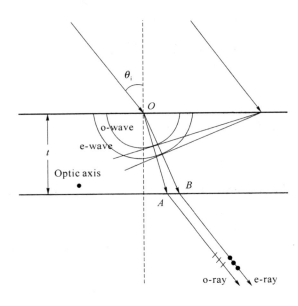

Chapter 15

1. (A) 2. (D) 3. (D) 4. (C) 5. (D) 6. (B)

7. **Answer:** 5.05×10^{-6} s; 1.36 km

8. **Answer:** 20 m; 19 m; $0.312c$

9. **Answer:** $0.806c$

10. **Answer:** $0.285c$

11. **Answer:** 4.19×10^9 kg/s

12. **Answer:** 4.50×10^{-10} J; 1.94×10^{-18} kg · m/s; $0.968c$

13. Answer:

$$\gamma = \frac{1}{\sqrt{1-\frac{v^2}{c^2}}} = \frac{1}{\sqrt{1-0.995^2}} = 10.0$$

In the reference frame S_1, $L_1 = 2.00$ m and $\theta_1 = 30.0°$.

Thus, $\quad L_{1x} = L_1 \cos\theta_1 = 1.73$ m, $L_{1y} = L_1 \sin\theta_1 = 1.00$ m

Therefore, in the reference frame S_2,

$$L_{2x} = \gamma L_{1x} = 17.3 \text{ m}, \quad L_{2y} = L_{1y} = 1.00 \text{ m}$$

(1) The proper length of the rod $L_2 = \sqrt{(L_{2x})^2 + (L_{2y})^2} = 17.4$ m.

(2) The orientation angle in the proper frame $\theta_2 = \arctan\dfrac{L_{2y}}{L_{2x}} = 3.30°$.

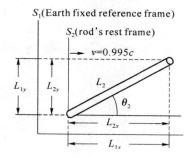

14. Answer:

Let frame S be the Earth frame of reference. Then $v = -0.7c$. The components of the velocity of the first spacecraft are

$$u_x = (0.6c)\cos 50° = 0.386c, \quad u_y = (0.6c)\sin 50° = 0.459c$$

As measured from the S' frame of the second spacecraft,

$$u'_x = \frac{u_x - v}{1 - u_x v/c^2} = 0.855c, \quad u'_y = \frac{u_y}{\gamma(1 - u_x v/c^2)} = 0.258c$$

Therefore, the magnitude of the velocity of the first spacecraft is

$$u = \sqrt{(u'_x)^2 + (u'_y)^2} = 0.893c$$

and its direction is at $\arctan\dfrac{u'_y}{u'_x} = 16.8°$ above the x'-axis.

15. Answer:

Conservation of both total momentum and mass-energy for the system gives

$$p_\nu = p_\mu = \gamma m_\mu u = \gamma(207 m_e)u \text{ and } \gamma m_\mu c^2 + p_\nu c = m_\pi c^2$$

From these equations and $\gamma = \dfrac{1}{\sqrt{1 - u^2/c^2}}$, we obtain

$$\gamma = \frac{1}{\sqrt{1-(0.270)^2}}$$

Then, the kinetic energy of the muon is

$$K_\mu = (\gamma - 1)m_\mu c^2 = (\gamma - 1)207(m_e c^2) = (\gamma - 1)207(0.511 \text{ MeV}) = 4.08 \text{ MeV}$$

The energy of the antineutrino is

Answers to Questions and Problems

$$E_\nu = m_\pi c^2 - \gamma m_\mu c^2 = (273 - 207\gamma)(m_e c^2) = 29.6 \text{ MeV}$$

Chapter 16

1. (D)   2. (B)   3. (B)   4. (B)   5. (D)
6. (D)   7. (B)   8. (A)   9. (D)   10. (D)

11. **Answer:** 264 nm; 4.70 eV

12. **Answer:** 97.3 nm; 3.08×10^{15} Hz

13. **Answer:** 1.7 T; 0.58

14. **Answer:** 57.9 m/s

15. **Answer:** $n\lambda/4, n=1,3,5,\cdots; n\lambda/2, n=0,1,2,\cdots$

16. **Answer:**

(1) $$\lambda = \frac{hc}{eV} = \frac{(6.63 \times 10^{-34} \text{ J·s})(3.00 \times 10^8 \text{ m/s})}{(1.60 \times 10^{-19} \text{ C})(18 \times 10^3 \text{ V})} = 6.91 \times 10^{-11} \text{ m}$$

(2) $$\lambda' = \lambda + \frac{h}{mc}(1 - \cos 45°) = 6.98 \times 10^{-11} \text{ m}$$

(3) $$E = \frac{hc}{\lambda'} = 17.8 \text{ keV}$$

17. **Answer:**

(1) The atom must be given an amount of energy

$$E_3 - E_1 = -(13.6 \text{ eV})\left(\frac{1}{3^2} - \frac{1}{1^2}\right) = 12.1 \text{ eV}$$

(2) There are three possible transitions

$$n=3 \to n=1, \Delta E = 12.1 \text{ eV and } \lambda = \frac{hc}{\Delta E} = 103 \text{ nm}$$

$$n=3 \to n=2, \Delta E = 1.89 \text{ eV and } \lambda = 657 \text{ nm}$$

$$n=2 \to n=1, \Delta E = 10.2 \text{ eV and } \lambda = 122 \text{ nm}$$

18. **Answer:**

$$\Delta E = -(13.6 \text{ eV})\left(\frac{1}{4^2} - \frac{1}{1^2}\right) = 12.8 \text{ eV}$$

$$\lambda = \frac{hc}{\Delta E} = 9.73 \times 10^{-8} \text{ m}$$

From Wien's law $\quad T = \dfrac{0.002,90 \text{ m·k}}{\lambda} = 2.98 \times 10^4 \text{ K}$

19. **Answer:**

$$\frac{\sin\theta_a}{\sin\theta_p} = \frac{1.22\lambda_a/D}{1.22\lambda_p/D} = \frac{\lambda_a}{\lambda_p} = \frac{h/p_a}{h/p_p} = \frac{p_p}{p_a} = \frac{\sqrt{2m_p K_p}}{\sqrt{2m_a K_a}}, K_a = 2K_p$$

$$\frac{\sin\theta_a}{\sin\theta_p} = \frac{\sqrt{m_p}}{\sqrt{2m_a}}, \sin\theta_a = \sqrt{\frac{1.67 \times 10^{-27} \text{ kg}}{2(6.64 \times 10^{-27} \text{ kg})}} \sin 15°, \theta_a = 5.3°$$

20. **Answer:**

(1) $$P = |\psi|^2 dV = (A^2 e^{-2\alpha r^2})(4\pi r^2 dr) = 4\pi A^2 r^2 e^{-2\alpha r^2} dr$$

(2) P is maximum where $\dfrac{dP}{dr}=0$. From $\dfrac{d}{dr}(r^2 e^{-2\alpha r^2})=0$, we obtain
$$2r-4\alpha r^3=0$$
$r=0$ corresponds to a minimum and $r=1/\sqrt{2\alpha}$ corresponds to a maximum.

At $r=0$, $|\psi|^2=A^2 e^{-2\alpha r^2}$ has a maximum, but the volume element $dV=4\pi r^2 dr$ is zero here, so P does not have a maximum at $r=0$.